The Illustrated Guide To Astronomy

THE ILLUSTRATED GUIDE TO ASTRONOMY

BILL YENNE

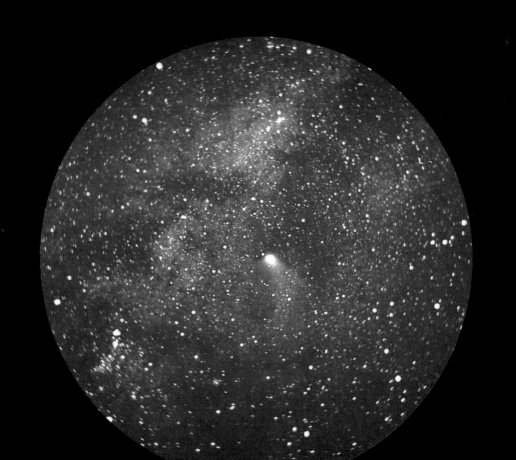

Published by
CHARTWELL BOOKS, INC.
A Division of BOOK SALES, INC.
110 Enterprise Avenue
Secaucus, New Jersey 07094

Produced by
Brompton Books Corp.
15 Sherwood Place
Greenwich, CT 06850

Copyright © 1993 Brompton Books Corp.

All rights reserved. No part of this publication may be reproduced, stored in a retrieval system or transmitted in any form by any means, electronic, mechanical, photocopying or otherwise, without first obtaining the written permission of the copyright owner.

ISBN 1-55521-959-4

Printed in Hong Kong

Designed by Tom Debolski
Captioned by Lynne Piade

Page 1: **The Trifid Nebula in Sagittarius, M20, NGC 6514.** *Page 2:* **Sunset at Kitt Peak National Observatory, as seen through the open dome of the Mayall four-meter telescope, one of the largest telescopes in the world.** *Page 3:* **The Comet Halley** *(center)* **passes through the Milky Way on the night of 14 April 1986, as photographed by Chile's Cerro Tololo Inter-American Observatory optical telescope.** *Below:* **A diagram of the Coudé Optical System used by selenographers in charting the Moon.**

ACKNOWLEDGMENTS

The author wishes to acknowledge Daniel Brocious of the Fred Lawrence Whipple Observatory in Amando, Arizona for the valuable assistance and information he provided during the preparation of this book.

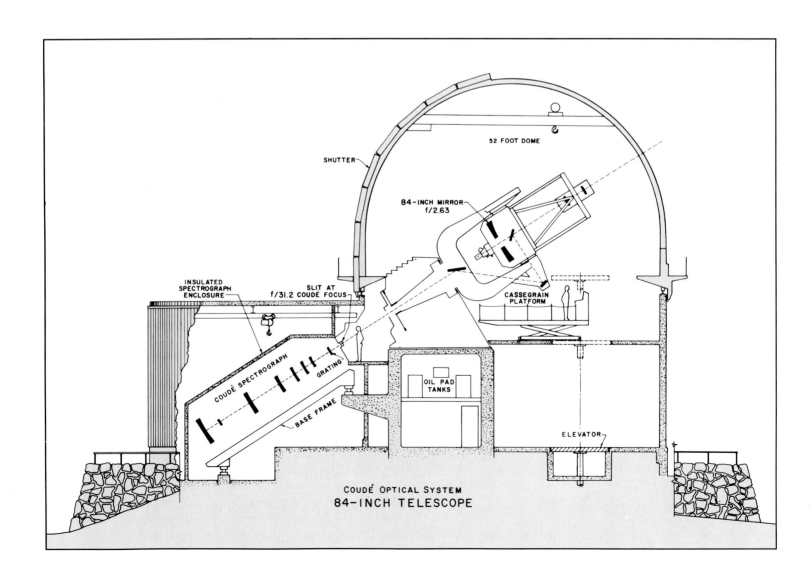

CONTENTS

INTRODUCTION 6

EARLY ASTRONOMICAL DISCOVERIES 10
Sidebar: The Pythagorean Theorem 12
Sidebar: Johann Kepler's Three Laws of Planetary Motion 18
Sidebar: The Origin of the Astronomical Telescope 19
Sidebar: Sir Isaac Newton's Three Laws of Motion 21
Sidebar: Transits of Venus 23

THE DEVELOPMENT OF MODERN ASTRONOMY 24
Sidebar: The Development of Astronomical Optics 27
Sidebar: The Discovery of Spiral Nebulae 28
Sidebar: The Discovery of Neptune 28
Sidebar: Selenographic Developments 30
Sidebar: Bode's Law 32
Sidebar: The Asteroids 34

THE SCIENCE OF ASTROPHYSICS 36
Sidebar: Progress in Accuracy of Measurement 45
Sidebar: The Discovery of Pluto 48

THE GREAT OBSERVATORIES 50
Sidebar: Astronomical Societies 53

STELLAR ASTRONOMY 58
Sidebar: Observing Novae 64

SOLAR ASTRONOMY 70
Sidebar: Values of Sun's Distance Current at Various Times 76

NASA SPACE SCIENCE 82
Sidebar: OAO 2 Discoveries 89
Sidebar: Observations Made by OAO 3 *(Copernicus)* 90
Sidebar: Mariner 5 Venus Observations 96

THE VIKING, VENERA AND MAGELLAN PROJECTS 102
Sidebar: Sir William Herschel's Observations of Mars 104
Sidebar: Discoveries Made by Viking Spacecraft 106

THE VOYAGER PROJECT 116
Sidebar: A Portrait of Jupiter 124
Sidebar: A Portrait of Saturn 134
Sidebar: A Portrait of Uranus 144
Sidebar: A Portrait of Neptune 148

A NEW GENERATION OF INTERPLANETARY SPACECRAFT 150
Sidebar: Third Generation NASA Interplanetary Explorers 150

THE NEW ORBITING OBSERVATORIES 150

YESTERDAY AND TOMORROW 166

APPENDICES 180

INDEX 189

INTRODUCTION

With our naked eyes, we can see about a thousand of the billions of stars that exist in a finite corner of the infinite universe. Many millions of years ago, human beings like us observed the stars and noticed patterns in their groupings. When they were outdoors at night, they would watch the stars and make up stories about them. People have always enjoyed telling stories and concocting myths to explain what they don't understand, and in this way, mythology was born.

Somewhere in the course of studying the heavens, the idea evolved that the configurations and movements of the celestial bodies truly affected human destinies and were 'signs in the heavens' from which future events could be foretold. If a comet appeared or an eclipse occurred at the time of the death of a great king or national leader, or when simultaneously with a remarkable celestial event, some plague, flood or drought afflicted the people, these sky and earth happenings were apt to be regarded as causes and effects.

The early people watched the stars on clear nights, imagining a godlike power there, and they picked out certain groups of stars, which they named after the heroes of their stories, and after familiar animals. Just as it is often easy, with a little imagination, to envision the forms of animals in clouds, so it was that early people imagined that certain star groups also looked like animals or other well-known objects. For example, humans found themselves competing with saber-toothed tigers and cave bears for food, so it is not hard to understand why the most prominent of constellations—Ursa Major—was named for a bear!

None of the primitive cave artists of Neanderthal Europe drew pictures of the stars of heaven, but ancient, civilized men did design pyramids 3000 years ago in Egypt and Mexico so that they were oriented toward Polaris, the North Star. Elements of what we now know as classical mythology traveled to ancient Greece and Rome from India, Chaldea and Egypt. Ultimately, the planets and constellations were named for Greek and Roman gods—who were nearly identical but called by different names—and meritorious animals found their shapes in the sky as rewards for services to the gods. Heroes were also designated, and designations within the heavens came to mirror the legends of Greeks and Romans.

The origin of Astronomy and astrology had common roots. The early evolution of Astronomy was devoted to the problems of the determination of the seasons and the calendar. Today, astrology is generally taken to mean the prediction of the fate of a person, as determined by the configuration of the planets, Sun and Moon, at the moment of his birth. This form is comparatively recent, and was preceded by a more general kind of prediction, frequently called **judicial astrology**, in contrast to the **horoscopic astrology**, which was for the individual. In judicial astrology, the celestial phenomena were used to predict the immediate future of a country or its government. From planetary configurations, eclipses, lunar halos, etc, the conditions of a coming harvest, flood, storm and so on were predicted, but the horoscope of a given individual was also involved. In other words, it was an elaborate omen prognostication. Horoscopes were not possible until the invention and use of methods for determining the positions in the sky of celestial bodies. That these early astrologers may have had some idea of scientific method is suggested by the report that the Babylonians tried to correlate horoscopes with the recorded events in the lives of individuals over a long period of time.

Astronomers of later days were much indebted in their investigations to the observations made and recorded by

Below: **Halley's Comet.** *Opposite:* **The star trails around Polaris mirror the rotation of the Earth in this view of Kitt Peak.**

Above: **The theologian and the astronomer discuss the universe.** *Opposite:* **Ptolemy's Earth-centered map of the universe.**

these primitive astrologer-astronomers, but the astrological predictions themselves depended on purely astronomical work of still earlier observers. These records have been useful from the time of the first star catalogue compilers **Hipparchus (c 150 BC)** and **Ptolemy (100-170 AD)**, up to the present century in such investigations as those of Cowell on the lunar orbit, who used records of solar eclipses in 1062 BC and 762 BC found on Chaldean baked bricks, and the tracing back of the recorded appearance of Halley's comet to 240 BC from Chinese observations.

The first phenomena to be noted would be the regularly recurring dawn, sunrise, daylight, sunset, twilight and night. Next to the measurement of a day thus provided, the month would be instituted as related to the variation of light with the Moon's phases. In temperate regions, where the first astronomical observations were systematically made, the changing length of the day, or the direction of the Sun at rising or setting, or the lengths of shadows cast at midday, would show that the Sun's daily path in the sky altered throughout the year, a time interval which was already marked by changing vegetation.

According to Sir WC Dampier, attempts were made to determine the number of months in the cycle of the seasons in Babylonia about 4000 BC and in China soon after. About 2000 BC, the Babylonian year settled down to one of 360 days or 12 months, the necessary adjustments being made from time to time by the interposition of extra months.

The discovery of a more precise length for seasons and a year was the work of a more specialized class of men than the primitive farmer or herdsman. This class was the priesthood or its equivalent. Before long, they found that there was not an exact number of months in the year or days in the month and had to undertake careful observations of the Moon and stars to obtain greater accuracy.

It seems likely that before this division had been effected, it would be found that the Moon in going like the Sun round the heavens always in the same direction from west to east (ie, opposite to the diurnal motion which she shares with the other bodies), kept in general to the same track in the sky. After a time, however, it would be noted by careful observers that this path was not constant, but deviated from the center line of the Zodiac, getting away from that line up to a maximum deviation on either side, but slowly returning to it. In the course of a number of years it would thus become evident that the Moon's path among the stars does not always lie in the same line on the celestial sphere, but in a zone or band about 20 moon breadths (10 degrees) wide, occupying the middle of the Zodiacal zone itself.

Among the brighter stars, Mercury, Venus, Mars, Jupiter and Saturn (the first two of which are never seen very far from the Sun in the sky) were soon noted to be moving in the Zodiac with varying periods. The name of **planet** (from the Greek, *planetes*, a wanderer) was later given to them because of their changing positions among the Zodiacal stars.

Before the Zodiacal belt was divided into **signs**, which according to some authorities took place about 700 BC when the intersection of the Ecliptic with the celestial Equator was in the constellation of Aries (the Ram), a number of asterisms, or configurations of stars in the sky, were arranged. The brighter stars of these configurations, thus identified, proved very useful in indicating the seasons of the year by their times of rising or setting. They were also useful for locating the positions on the celestial vault of moving objects such as planets, comets and shooting stars, and in helping the traveler by land or sea to determine direction. These configurations are called **constellations**.

In written history, a few constellations were mentioned in the Bible and by **Homer (c 800 BC)** and **Hesiod (c 700 BC)**, but the first complete description of the constellations was by **Eudoxus of Cnidus** in 366 BC. This work, which Aratus turned into verse in his *Phainomena* in 270 BC, is the chief, original source of lore about the heavens. The great Greek astronomer and astrologer Claudius Ptolemaes, aka **Ptolemy (100-170 AD)**, catalogued 1028 stars, and elaborated on the astrological nuances of the constellations and the legends about them previously described by Eudoxus and others. His book *Almagest* (150 AD) was the foundation of much later work on the subject.

Forty-eight named constellations have come down from extremely ancient times. Others, chiefly situated in southern skies, have been added in modern times. In many, the stars form a well-marked group, clearly separated from other groups, and the names given to these formations are supposed to have been suggested by a resemblance to the shapes of certain familiar objects. The resemblance is usually very slight, and this may justify the suspicion that often some fancied figure was first thought of, and then the stars were chosen to represent it in a very rough fashion.

Individual stars were named by Arab astronomers, but the Greco-Roman constellation names persist. In 1603, the astronomer Johann Bayer formulated a method for using the constellation names to identify stars. The brightest star in a constellation would be given a name that used 'Alpha' plus the name of the constellation. The second brightest used 'Beta,' the third, 'Gamma' and so on through the Greek alphabet. For example, the two brightest stars in the constellation Orion are known by their popular Arabic names as Betelgeuse and Rigel. However, under Bayer's system, they're known as Alpha Orionis and Beta Orionis.

The similarity of the constellations, as early recognized in different countries, is remarkably great. This points to a common origin for them. The late Dr ACD Crommelin thought that there was reason to believe the stars may have been grouped to some extent by the Egyptians as early as 4000 BC, and he remarked on their use of the then Pole Star for orienting the Great Pyramid. The Chinese also mapped the sky into many divisions of stars by 2500 BC.

Johann Bayer (1572-1625) also added constellations to the map by picking out fainter groups that hadn't been mentioned by earlier sky watchers. The telescope made such tasks much easier, and astronomers such as the Frenchman **Nicolas Louis de La Caille (1713-1762)** and the Prussian **Johannes Hevel (1611-1687)** followed Bayer's lead, identifying new constellations and naming them after objects—like chemical ovens—that had not existed in Ptolemy's day.

While the ancient Greeks and Semitic scholars working in Alexandria in Ptolemy's time gave Arabic names to individual stars and Bayer applied his own scheme to these same celestial friends, there were still objects in the sky to be reckoned with. These objects—galaxies, star clusters and nebulae—which, though fainter and vastly farther away, were larger than stars, were first catalogued by **Charles Messier (1730-1822)**, who identified them with a designation beginning with his initial 'M.' Today, the Messier objects tend to be closely associated with constellations. For example, the beautiful, and often photographed, galaxy M31 is more commonly referred to as the 'Andromeda Galaxy.'

The fact that some observed celestial objects are closer to the Earth than others must have shown itself to careful observers at a comparatively early date by the occurrence of eclipses and by occultations, which are caused by the passage of the Moon between Earth and a planet or a fixed star. From this, the Moon was soon noted to be nearer to us than the other heavenly bodies.

As an indication of the nearness of the planets and the Sun and Moon, the rapidity of their motions with respect to the fixed stars was taken. The three slower moving planets, Saturn, Jupiter and Mars, were thus considered to be beyond the Sun, and were called the **superior planets**. The other two planets, Venus and Mercury, which appear to accompany the Sun in its annual path in the sky, never very far from it, were termed the **inferior planets**. Eclipses of the Sun and of the Moon must have roused great interest, and often alarm. That the former always happened at New Moon and the latter at Full Moon was noticed, and the cycles of their occurrence would be noted after the accumulation of numerous records of dates of observation.

Other great discoveries, not confirmed until relatively recent times, are Earth's spherical shape and the diurnal motions of celestial bodies in rising in the east and traversing the sky to setting in the west. The early observers thought of the Earth as a circular plane over which the celestial vault extended, the hemispherical outline defining the extremities of the Earth in all directions. The Sun, Moon, and stars appeared to move on, or within, the inner surface of this vault. Just as the ocean was supposed to flow in a stream around the outer margin of the plane of the Earth, the heavenly bodies were believed to emerge from the ocean when they rose, and to sink into it when they set. The exceptions were **circumpolar objects**, which were near enough to the celestial pole to avoid this immersion. Other views seem to have been that the Sun did not pass under the Earth but that it traveled to the east after its setting around north, night being caused by an elevation of the northern part of the Earth which shut out the light of the Sun.

Nevertheless, in spite of these notions, there was perhaps a larger proportion of early mankind aware of the apparent movements of the heavenly bodies than there is today. It has indeed been remarked that as civilization progresses, mankind's interest in the actual observation of the heavens diminishes.

The orientation of ancient monuments like **Stonehenge** and the avenues of Standing Stones at Camac in Brittany, as well as of many other large megalithic monuments, demonstrates a clear knowledge of the Sun's position in the heavens with regard to the seasons. This could only have been acquired by careful astronomical observation. At Stonehenge, during the summer solstice, the Sun was observed to rise over a stone known as the Friar's Heel—which stands on the common axis of the two circles and two horseshoes of standing stones. The Sun could be viewed through the aperture of the arch formed by the two upright stones and the capping stone of the Great Trilithon. At Camac there are three huge avenues of standing stones. One of these avenues faces the east—the point of sunrise at the equinoxes—while the other two face the northeast—the point of sunrise at midsummer, or the summer solstice. Approximate dates for the setting up of such stones have been worked out, allowing for the alteration in the points of the sunrise caused by change in the Obliquity of the Ecliptic—the angle between the plane of the Earth's equator and the plane of the Earth's orbit round the Sun. Stonehenge dates from about 1800 BC, although archaeological research indicates that this monument is not all of one period but probably was gradually built over many years after that date. That time is of course contemporary with more advanced astronomical observation in countries of a higher degree of civilization such as Babylonia and Egypt.

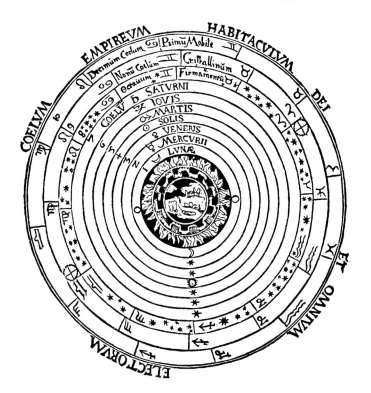

Early Astronomical Discoveries

The foundation of much of our current astronomical knowledge originated in ancient times. The Egyptians, for example, divided a zone of the sky into 36 small star groups, or 'dekans,' which by their heliacal risings marked the beginning of each 10-day period. It also appears that they had a value for the Obliquity of the Ecliptic before the Chinese measurement of 1100 BC, and they knew that the length of the year is about 365.25 days. They were the first to adopt the year instead of the month as the standard time measurement, and they employed 365 days for it. They have also been credited with a belief that Venus and Mercury revolve round the Sun.

Members of the Egyptian priesthood were the observers. They alone possessed the acquired knowledge of the subject. This privileged class performed the valuable function of predictors for the Nile floods, which irrigated and fertilized large parts of the country. The priests of Thebes claimed to be the originators of exact astronomical observation, attributing this partly to the clear skies of their country. They determined the periods of the planets around their circuits of the sky, but their assertion of a general priority in exact observation has not been proved.

Some authorities believe that the **gnomon** and **sundial** were first introduced in Egypt. A gnomon is a vertical shaft or column erected on a horizontal plane. The measurement of the length of the shadows cast by it provide the means of calculating angles of elevation of the Sun. Portable instruments for observing transits of stars across the meridian, for orientation of monuments and probably for time measurement, certainly existed at, and perhaps before, the time of Tutankhamon (c 1355 BC).

The astronomers of ancient Mesopotamia are credited with the invention of the **clepsydra**, or water clock. The Babylonians probably had this clock as early as the seventh century BC. It consisted of one vessel kept full of water that was allowed to escape through a small hole in the bottom into a receptacle, the rise of the level in the latter showing the passage of time by a pointer, on a float, directed to a graduated scale. It is said that in one instrument the amount which was allowed to pass in a day and night corresponded to about six drops per second. The gnomon and the sundial may also have been invented by the Babylonians. Both were instruments of the greatest value before the later improvements of graduated circles and mechanical clocks.

Celestial phenomena were observed and recorded in China more than four thousand years ago. One of the earliest of these refers to the making of a sphere, to represent the motions of the celestial bodies, by **Yu Shih** at a date shortly after the reign of the emperor Fuh Hsi about 2950 BC.

Chinese observations of comets—372 of which have been found in the records between 611 BC and 1621 AD—have been useful in modern times for identification of comets which return to the Sun's vicinity in a periodic orbit. In the case of Halley's Comet, it is thus known that it certainly was observed in 240 BC, and probably also in 467 BC. (Its average period is about 76 years.)

Halley's Comet was seen by the Chinese astronomers at a number of its returns since these dates, when no European record is available. One comet, seen by them in 134 BC, also appears in the first star catalogue by the Greek astronomer Hipparchus.

The early Chinese astronomers were evidently aware of the fact that the tails of comets point away from their heads in a

Opposite: **The Egyptian Sphinx.** *Above:* **The Pyramids on the Plain of Giza. The Egyptians were accomplished astronomers.**

direction opposite to that of the Sun. In his collection of observations, the astronomer Ma Tuan Lin noted, 'In general, in a comet east of the Sun, the tail, reckoning from the nucleus, is directed to the east, but if the comet appears to the west of the Sun, the tail is turned towards the west.'

The Hindus began to treat Astronomy as a true science in the 'Siddhantas,' a series of works which appeared between 300 AD and 1300 AD. In one, an astonishingly accurate diameter is given for the Earth: 1600 *yojans* of four to nine miles, or 7840 miles. A total of 51,570 yojans, or 253,000 miles, is given for the distance of the Moon. Rules for calculating positions of the heavenly bodies are found and a knowledge of the differences between a sidereal and solar day and a sidereal and solar year is shown in others.

The evidence all indicates that from about 400 AD, the Greek system of Astronomy was adopted by the Hindus, their astronomers having absorbed Greek views although such views were not accepted in Europe until much later. It may be that the time of introduction of the Greek ideas was much earlier, perhaps between the period of Hipparchus (c 150 BC) and Ptolemy (c 140 AD).

The earliest Greek scientific work was the product of Ionian colonists in Asia Minor, who occupied the coastal areas near Ephesus and Halicarnassus. They were a people very favorably situated for the reception of foreign ideas. To the east, they were in contact with current Mesopotamian culture, and their sea traffic with Egypt brought knowledge from the ancient civilization there. Moreover, prevailing political conditions favored intellectual progress. It may also be said that, although the Egyptians and Mesopotamians had gone further than the ancient Greeks in mathematics and in astronomical observation, the purpose of their studies were almost entirely utilitarian. The Ionians, and later the Greeks, took up these matters in a more purely intellectual manner and built the foundations for the science of Astronomy.

Thales (624-545 BC) of Miletus near Ephesus is commonly called the founder of Greek Astronomy. From the Mesopotamian and Egyptian astronomers, Thales knew the length of the year, the inequality in the length of the seasons, the positions in the sky of the solstices and equinoxes, and the signs of the Zodiac. Thales' pupil, **Anaximander (611-547 BC)**, also of Miletus, introduced the gnomon and sundial from Babylonia. He was familiar with the Obliquity of the Ecliptic and believed that the Earth's shape was that of a cylinder, with its height a third of its breadth. He appears to have been the first person to speculate on the relative distances of the heavenly bodies, and he supposed the heavens to be of a fiery nature, spherical in form, enclosing the atmosphere 'like the bark of a tree.'

The great philosopher **Plato (422-347 BC)**, although not an astronomical theorist, made several references to the subject at various places in his writings. In one of his dialogues, he gives a short account of the heavenly bodies, their arrangement and motions. Starting at the nearest body, the order he gave was: Moon, Sun, Mercury, Venus, Mars, Jupiter, Saturn,

Left: The shape of the Great Pyramid of Cheops at Giza has intrigued mathematicians for centuries after it was discovered that the ratio of its height to twice the base would equal pi, a number of universal significance. It is unknown whether the Pyramids were just oversized mausoleums or if they had any astronomical relevance. *Opposite:* Ptolemaeus, or Ptolemy, the father of western astronomy, is depicted with his model of the spherical Earth.

the stars. The outer planets moved more slowly, he believed, but Mercury and Venus performed their revolutions in the same time as the Sun. This appears to indicate that Plato was aware that the motions of Mercury and Venus were of a different type than those of the others. It is recorded that he set his pupils the problem of propounding rules by which the movements of the Sun, Moon and planets could be reduced to a combination of uniform circular or spherical motions.

Plato's legendary student **Aristotle (354-322 BC)** adopted the ideas of Eudoxus and Callipus, but he added 22 spheres to allow for what he thought were disturbing effects of the spheres on one another, thus increasing the total to 56. He treated them as material entities and the result was, from a mechanical point of view, a very confused conception. He believed the shapes of the heavens and the various celestial bodies to be spherical, basing this in the case of the Moon on its phases, which were possible only in a spherical body illuminated by the Sun. His writings summed up the state of astronomical knowledge and ideas of the period. To his mind, the Earth's rotundity was proved by the circularity of its shadow on the Moon during eclipses, and by the change in altitude of the stars in the sky as an observer moves north or south. However, he did not believe that the Earth revolved around the Sun, owing to the absence of any observed consequent displacement of the stars in the sky. He believed the Sun and Moon were nearer to the Earth than the planets because he had observed an occultation of Mars by the Moon and knew of similar observations by Egyptians and Babylonians. The ideas that Aristotle advanced, although not original, were in substance and by their symmetry, good enough to last for more than 2000 years.

Probably the greatest astronomer of ancient times was **Hipparchus (190-120 BC)** of Nicea. He does not appear to have belonged to the school of Alexandria but it is thought that he may have visited and observed the stars there. Hipparchus may be regarded as the founder of systematic observational Astronomy. Greatly developing the study of trigonometry and its application to astronomical problems, he made a very large number of observations, and collected and collated records of the work of previous observers in order to discover, if possible, any astronomical changes which might have taken place. These earlier observations extended back beyond those of the Alexandrian School and older Greeks, and included some of the still more ancient records of Babylonia. He did the greater part of his work at an observatory which he built on the island of Rhodes.

In comparing his observations with those of Aristyllus and Timocharis about 150 years before, Hipparchus found that there had been changes in the distances on the sky of certain stars from the two points of intersection of the circle of the Ecliptic and the celestial Equator, known as the points of the equinoxes. These changes were of a kind only to be explained by a motion of the equinoxes in the direction of the apparent daily movement of the stars east to west, and not by actual movement on the sky of the stars concerned. Thus Spica, the chief star in the constellation of Virgo, had become further separated from the Autumnal Equinox by an angle of two degrees in 150 years, or 48 seconds of arc per year. The correct amount of this movement is 50.3 seconds annually. It is due to the revolution of the Earth's pole around the pole of the Ecliptic in a period of 26,000 years in a direction opposite to

THE PYTHAGOREAN THEOREM

Pythagoras (582-500 BC) was one of the Western world's first great mathematicians, and originated the **Pythagorean Theorem**, which stated that the 'square of the hypotenuse of a right triangle is equal to the sum of the squares of the other two sides' (A+B=C). Pythagoras also held that the Earth, Moon, planets and fixed stars all revolved around the Sun, which itself revolved around an imaginary central fire. The Pythagoreans seem to have been the first to maintain that the Earth and other heavenly bodies were spheres. They appear to have arrived at their conclusions from views that the sphere was the perfect shape, circular motions the most perfect motions, and that 10 (which=1+2+3+4) was the perfect number (number being the real substance of all things). Pythagoras also theorized that Phosphorus and Hesperus, the morning and evening stars, were one and the same body (Venus).

that of the Earth's revolution around the Sun. A consequence is that the Sun, in its annual journey around the sky in the ecliptic, returns to each equinoctial point a little earlier each year with respect to the stars, so that the equinoxes occur slightly sooner than they otherwise would. The movement has thus received the name, Precession of the Equinoxes.

Hipparchus also measured the Obliquity of the Ecliptic, arriving at the same result as Eratosthenes. An important discovery made by comparison of his own observations with those of Aristarchus was that the **Tropical Year**—the time enclosed by two successive passages of the Sun through the Spring Equinox, or 365 days—hitherto accepted, was several minutes too long.

Appolonius of Perga (c 250-200 BC) had suggested that the motions of the heavenly bodies can be represented more simply and accurately by combinations of circular motions than by the revolving spheres of Eudoxus. When Hipparchus examined the movements of the planets, he therefore considered two theories of the circular motion—that of **eccentrics** and that of **epicycles**. Some of his predecessors had adopted the latter, that is to say, they had held that each planet moved uniformly on a smaller circle (an epicycle), the center of which itself moved uniformly upon a greater circle (known as the **deferent**) at the center of which is the Earth.

In his **Eccentrics Theory**, he assumed that in the case of the Sun, for example, the center about which it was supposed to move in a circular orbit round the Earth was situated at a little distance from the Earth. Hipparchus finally adopted a systematic order of eccentrics, and to a lesser extent, of epicycles. He developed this for the Sun and Moon with considerable success. However, his attempts with the planets were not followed up, owing in large part to inadequate numbers of satisfactory observations, and contented himself with systematically collecting observations for use by his successors. Hipparchus' methods, and the tables constructed by their means, enabled his successors to predict the times of eclipses of the Moon within an hour or two and eclipses of the Sun somewhat less closely, both much more accurate than previously possible.

Hipparchus also noted that the seasons were unequal. The time between the Vernal Equinox and Summer Solstice is 186 days, but from the Summer Solstice to the next Vernal Equinox is only 178.75 days. He measured the inclination of the Moon's orbit to the Ecliptic as five degrees. He made an estimate of the size and distance of the Moon by an eclipse method. By observing the angular diameter of the Earth's shadow on the Moon, and comparing it with the known angular diameters of the Sun and Moon, he calculated a relation between the distances of the Sun and Moon, from which either distance could be found when the other was known. He knew that the distance of the Sun was much greater than that of the Moon, and trying more than one distance for it, he obtained results showing that the Moon's distance was nearly 59 times the Earth's radius. Combining this finding with Aristarchus' estimate that the Sun's distance was 19 times that of the Moon, a distance for the Sun of about 1100 times the Earth's radius was obtained. This was more than 20 times too small, the true distance being about 23,400 times that unit. He found the size of the Moon to be a quarter of the diameter of the Earth.

An invention of the greatest value to observational Astronomy, probably due to Hipparchus, was the **astrolabe**, an instrument for finding the latitude and longitude of a celestial body relative to another body for which these coordinates

are known. **Celestial latitude** is the angular distance North or South of the ecliptic. **Celestial longitude** is the arc of the ecliptic between the Vernal Equinox (First Point of Aries) and the foot of a great circle drawn from the pole of the ecliptic, through the object, to the ecliptic.

No astronomer who preceded him had a greater influence on the history of Astronomy than **Claudius Ptolemaeus**, or **Ptolemy(c 140 AD)**, who lived and worked near Alexandria from 127 AD to 151 AD. His chief written work is *Mathematike Syntaxis*, or the *Almagest* (an Arabic title derived from the Greek, meaning The Greatest). In this book, based on the results of previous astronomers, especially those of Hipparchus, Ptolemy adopted the hypothesis of a fixed spherical Earth and assumed that it was the largest of the heavenly bodies, although merely a point in comparison with the distance of the fixed stars.

In the Ptolmeic system, a planet revolves uniformly in a circle (**epicycle**), the center of which revolves uniformly on another circle (the **deferent**) around a point not coincident with the Earth (**eccentric**). A fair approximation to a correct representation of the planetary motions is arrived at by adjusting the proportions of the radii of these two circles, the velocity of the planet and the eccentric.

During the five centuries after the death of Ptolemy, the Alexandrian school continued to exist, but very little was produced in the way of writings in Astronomy except a few works by commentators and compilers. In fact, it may be said that the history of the Astronomy of Greece and Alexandria practically stopped after Ptolemy. Observations seem to have been abandoned, so much so that only eight are known to

Above: The Persian astronomer Avicenna (980-1037). *Below:* In an illuminated manuscript, Christian kings and philosophers of antiquity observe the heavens. The planets Mercury, Venus, Mars, Jupiter and Saturn are depicted as stars. *Right:* The Great Caracole at Chichen Itza, an ancient Aztec observatory in the Yucatan peninsula. Much of the Aztecs' knowledge of astronomy was appropriated from the Maya, whose astronomers were charting the movement of the planets 4000 years ago.

have been recorded until the time of the Arabian astronomers in the ninth century. Perhaps in no other science more than Astronomy are the Middle Ages (c 500 to 1500 AD) in Europe most aptly described by the appellation 'Dark' Ages. However, as we know, an extraordinary revival of interest in science generally, and particularly in Astronomy, took place among the Arabs whose ascendance coincided with the decline of intellectual development in Europe. Ptolemy's works were translated several times into Arabic during the ninth century, and several other Greek works were similarly treated.

While Arab astronomers cannot be said to have advanced any original theories, they were very accurate and painstaking observers, and competent calculators. The introduction of the decimal notation with a consequent enormous simplification of arithmetic was due to the Arabs, and science also owes to them the names of many of the brightest stars, and of astronomical terms such as 'almanac,' 'zenith' and 'nadir.'

Perhaps the greatest service rendered by the astronomical achievements of the Arabs from the eighth to the fifteenth century is the prevention of any total break in the cultivation of Astronomy.

With the revival in classic learning, Ptolemy's works, previously known only in poor Latin translations from Arabic versions themselves not faultless, began to be read in the original Greek, rousing fresh enthusiasm. **George Purbach (1423-1461)** and his extremely able pupil **Johannes Muller (Regiomontanus) (1436-1476)** of Konigsberg, Franconia were in succession professors of Astronomy and mathematics at Vienna, where they applied themselves to the examination and improvement of the Ptolemaic system.

Leonardo da Vinci (1452-1519), the universal genius who is celebrated for his knowledge and proficiency in all the arts and sciences, wrote of the rotation of the Earth as being common knowledge. He was the first astronomer to explain the illumination seen on the Moon's disc inside its crescent when it is very narrow, as the effect of sunlight reflected onto it from the Earth. This phenomenon is known as the **Earthshine on the Moon**—a 'reflection of a reflection.' The less the Moon appears illuminated by the Sun from the Earth, ie, the narrower the crescent, the more illuminated is the Earth as seen from the Moon. The Earth is *full* viewed from the Moon, when the Moon is *new* to us, and vice versa.

A new spirit of inquiry marked the beginning of the sixteenth century whereby the authority of Ptolemy, and even of Aristotle, was actively questioned. To this inquiring generation belonged the great Polish astronomer **Niklas Koppernigk (1473-1543)**, or as he later became known, **Nicholas Copernicus**. Although he published no important work until the latter part of his life, he was well-known in the intellectual circles of Europe as an astronomer and mathematician. A complete account of his studies and their results was not made public until just before his death. In fact it has been said that a copy of his newly printed book *De Revolutionibus Orbium Celestium* was given to him on his death bed, but probably never opened by him.

The first of his theories concerned the rotation of the Earth. Previously, the rising, passing across the sky, and setting of all but circumpolar objects had been largely attributed to a real movement of the bodies themselves. Copernicus saw that the phenomena could be explained either in that way or by the Earth's rotation *around an axis* in the opposite direction. Although Ptolemy had noted the great difficulty involved in the former explanation due to the tremendous velocities of

the stars which it entailed, he had accepted it rather than the latter explanation because of the absence of movements of loose objects or of the atmosphere which he thought would take place if the Earth was actually rotating. However, Copernicus realized that the atmosphere and any loose bodies must accompany the Earth's surface in the rotation supposed, having naturally the same movement themselves. He adopted the idea of the rotation of the Earth as correct, believing it to be unlikely that the enormously greater universe should be in rotation instead.

Copernicus retained the old idea of uniform circular motion and had to assume different centers for planetary orbits, outside the Sun. Epicycles were also necessary, as they had been with Ptolemy's system. Several epicycles had to be introduced, one upon the other, to explain observed irregularities in the Moon's orbit. The old epicyclic theory was simplified only in the placing of the center of the Earth's orbit at the center of the universe. The movements of the planets were also referred to this point through which the planes of their orbits passed, and their positions of greatest and least orbital velocity were related to it. The Sun was placed in a position near to the center of the Solar System, but, as Copernicus wrote, 'did not *seem* to have any physical connection with the planets as the center of their motions.'

Copernicus was fully aware that the revolution of the Earth in its orbit must cause apparent displacements of the stars in the sky (**parallactic displacements**), but he considered the stars' distances from Earth were so great as to make it impossible to detect the very small displacements.

The Copernican system was not accepted as a true one by all scientists of the time. For example, Francis Bacon showed no real appreciation and referred to Copernicus as a 'man who thinks nothing of introducing fiction of any kind into nature, provided his calculations turn out well.' Among astronomers, however, his work was gradually adopted as a great advance, and today we view it as the turning point in mankind's conception of the universe and Earth's place in it.

Tycho Brahe (1546-1601), the great Danish astronomer, first became interested in science after witnessing a partial eclipse of the Sun as a boy on 21 October 1560. He bought a copy of Ptolemy's works, which is still in existence with his marginal notes in the Prague University library.

Along with friends, Tycho built several astronomical instruments in Germany, including a huge wooden quadrant with a radius of 19 feet. In 1570, he returned to Denmark, where in 1572 he observed the supernova in Cassiopeia. It was this astronomical event which probably finally fixed his life interest in Astronomy. He wrote the book *De Stella Nova* in which he recorded his systematic observations of the star's changes in brightness and color, its position in the sky, and the absence of movement which showed that it indeed belonged to the distant fixed stars.

16 THE ILLUSTRATED GUIDE TO ASTRONOMY

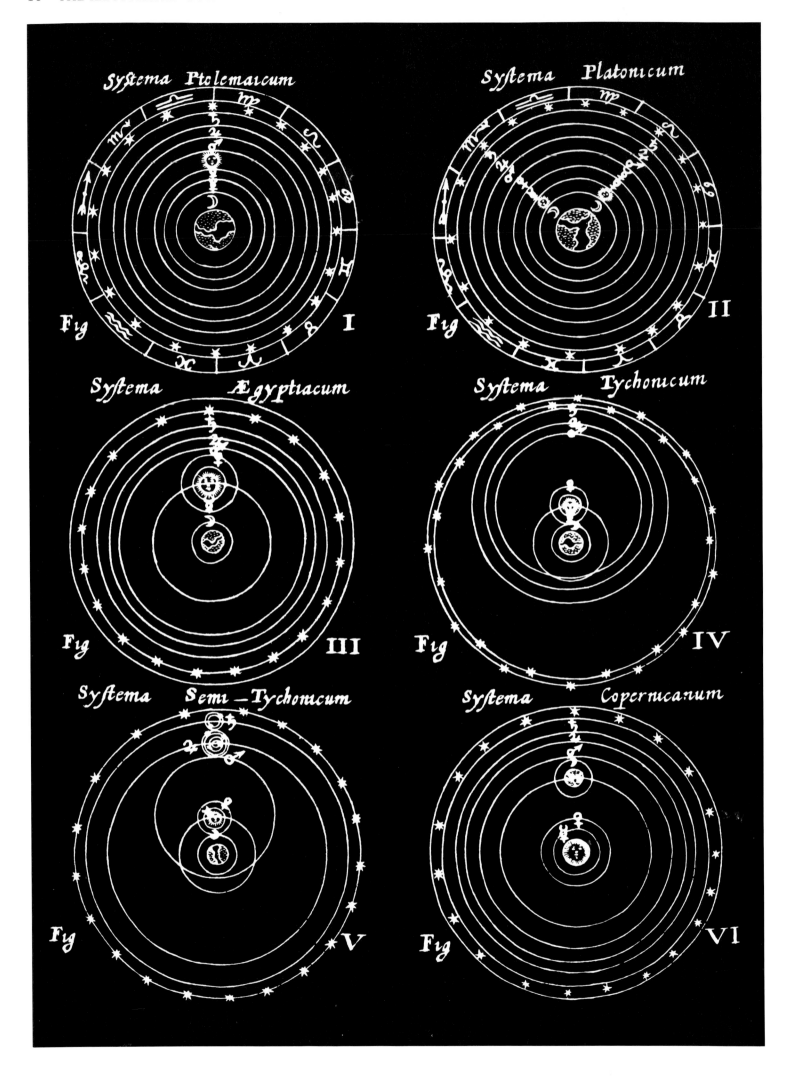

By using Tycho's refined and improved methods of observation, much greater accuracy was obtained as compared with the results of former astronomers. This accomplishment is all the more remarkable given that Tycho had no good mechanical clocks and had to resort to using clepsydra, with mercury instead of water, to measure time. To allow for this deficiency in his apparatus, he sometimes used procedures which were not dependent on clocks. In 1582, for example, he made use of the fact that the planet Venus was six weeks visible in daylight even before it crossed the meridian, and he could therefore measure its angular distance from the Sun and from bright stars.

Among Tycho's outstanding achievements were the fixing of the position of a star used as a standard by astronomers (Hamal, Alpha, Arietis) within a quarter of a minute of arc of its true place, with a value for Precession of 51 minutes. He discovered some irregularities in the Moon's motion, previously unknown, and calculated the inclination of the Moon's orbit to the ecliptic oscillates in value from about five degrees to 5.33 degrees.

Tycho also compiled a catalogue of 977 of the fixed stars, the positions of 200 of which were, however, not up to the accuracy of the others. He wrote a book on comets, which proved that they were celestial bodies, further away than the Moon and planets, and not appearances in the Earth's atmosphere.

Tycho therefore devised a system of his own in which the planets moved round the Sun, but the Sun—with its planets—and the Moon, revolved around the Earth, a system similar to that of Julianus 1200 years before. There were still some epicycles and deferents (the Sun's orbit was the deferent of the planets' epicycles) necessary to allow for irregularities, which made his system complicated in much the same way as those of Ptolemy and Copernicus.

Johannes Kepler (1571-1630) was one of the key figures in what we might well describe as the dawn of modern Astronomy which itself took place at the dawn of the seventeenth century. By 1600, Kepler was working with Tycho Brahe. Tycho introduced him to Emperor Rudolf, who conferred on him the title of Imperial Mathematician, on the condition that he assist Tycho in his calculations. When Tycho died in 1601, Kepler was appointed as his successor, but he was shabbily treated in that his salary was only half that given to Tycho. Not being an observer, he never acquired any of Tycho's instruments, but he secured control of the greater part of the records of the magnificent series of observations that Tycho had made, chiefly at Hven. For the next 25 years of Kepler's life, these observations became the basis of an improved theory of the Solar System and of several epoch-making discoveries.

Carefully examining a number of hypotheses—such as those which stated that the planets revolved around centers outside but near to the Sun and that the centers of epicycles are not situated exactly on deferents—he found discrepancies in position amounting to as much as eight minutes of arc. These, he declared, were impossibly great for the observations of Tycho which he was utilizing. He also compared the Ptolemaic and Tychonic systems with observation and found them defective as well.

After a great deal of computation, adopting many hypotheses, Kepler advanced his **Three Laws of Planetary Motion**.

The first two of these Laws were published in his book *Astronomia Nova,* in 1609, but it would take nine years longer

Left: **Six models of understanding the planetary system as devised by Ptolemy, Plato, the Egyptians, Tycho Brahe, an unnamed student of Brahe and Copernicus.** *Above:* **Tycho Brahe postulated that the planets revolved around the Sun which in turn revolved around the Earth.** *Below:* **Copernicus was the first astronomer to assert a heliocentric system.**

Above: **Johannes Kepler**, whose Three Laws of Planetary Motion laid the foundation for Newton's discovery of the Law of Gravity. *Right:* A diagram of Tycho Brahe's astrolabe.

to discover the third. It was later demonstrated by Sir Isaac Newton that these Three Laws precisely defined the conditions under which planetary revolution must proceed if governed by a force emanating from the Sun and decreasing as the square of the distance from that body increases, ie, the law of gravitation as exemplified by the Sun and its planets.

As well as these major discoveries, Kepler made many other valuable contributions. He drew attention to the use of eclipses for determining differences of terrestrial longitudes (the differences between local times of occurrence of the eclipses gives this). He also speculated on the physical causes behind his Three Laws and suggested that these were attractions between the Sun and planets which varied with distance and were proportional to mass. He regarded tides as being due to a mutual attraction between the Moon and the seas of the Earth. However, the state of the mechanical ideas of his time was too undeveloped and imperfect for greater success in such speculations.

Kepler also wrote on optics, and was the first to suggest that a telescope, with both object glass and eyepiece convex lenses could have cross wires fitted in the focus to help to accurately fix the positions of stars seen in its field.

The great Italian astronomer and physicist **Galileo Galilei (1564-1642)** was born at Pisa, the son of the noted mathematician, Vincenzo Galilei. His contributions to astronomy—particularly observational astronomy—were enormous. His important astronomical work probably began with his observations of the New Star in Ophiuchus found by Mastlin in 1604. This star, which attained a brightness nearly equal to that of Venus, was seen for about 15 months before fading to invisibility with the naked eye. From the absence of parallax, he disproved the commonly-held idea that it was some type of meteor, and that, like the fixed stars, it was situated beyond the bounds of our own Solar System.

In January 1610, Galileo achieved the momentous discovery of moons in orbit around a planet other than the Earth. Turning his 32-power telescope to Jupiter, he noticed the four tiny points of light that we now know as Io, Ganymede, Europa and Callisto. These four largest moons of the largest planet are known collectively as the **Galileans**.

Looking beyond the planets, Galileo located 36 stars in the Pleiades, where ordinary eyesight shows only six or seven and 40 stars in the cluster Praesepe, where previously only three nebulous stars were thought to exist. Under telescopic scrutiny, he found that the Milky Way was composed of myriads of stars.

Galileo postulated that the determination of stellar distances from displacements in the sky could be made with

JOHANNES KEPLER'S THREE LAWS OF PLANETARY MOTION

1. A planet travels in an orbit which is an **ellipse**, and the Sun is in one of the two foci. It is worth noting that in 1080 Arzachel, and Reinhold in the early sixteenth century, had suggested an oval or elliptical form.
2. The **Law of Area**: The planet travels in its orbit at a rate such that the **radius vector**—or line joining it to the Sun—sweeps out equal areas in equal times.
3. The **Harmonic Law**: The cubes of the mean distances of the planets from the Sun are proportional to the squares of the periods of revolution.

THE ORIGIN OF THE ASTRONOMICAL TELESCOPE

In 1609, Galileo had heard that a **Hans Lippershey** or **Lippersheim (?-1619)**, a maker of eyeglasses in Holland, had produced an instrument consisting of two lenses which magnified distant objects. Working with this information, he put two lenses into a tube, the one furthest from the eye (the object glass) being planoconvex, the eye-lens planoconcave. He made a number of these instruments, the largest magnifying 32 times.

Galileo was not the inventor of the telescope. This accomplishment has been variously attributed to Roger Bacon (c 1260), Leonard Digges (c 1558), Portia (c 1558) and to Jansen and Metius in the Netherlands. However, Galileo *is* in fact generally recognized as the first to use a telescope in Astronomy. While it seems possible that an Englishman, **Thomas Harriott (1560-1621)**, and a German, **Simon Mayer** or **Marius (1570-1624)** may have used a telescope shortly before he did, their work in the field was trifling compared with his.

On directing his telescope to the sky, Galileo was rewarded with a remarkable series of discoveries by which his name quickly became celebrated throughout Europe. These included his discovery of the spots on the Sun, which appeared to rotate around it in about 27 days, and mountains, craters and plains on the Moon. He also discovered that planets were discs of appreciable size and not points of light like the fixed stars. Venus showed phases like the Moon.

It is possible that Galileo's telescopic discoveries have been given rather too much prominence, as they were inevitable sooner or later once the telescope had been invented. However, he used these discoveries in support of the truth of the Copernican system, and his work on dynamic subjects leading to the recognition of the laws of motion and of force as the cause of motion was entirely new and independent, a brilliant example of the true scientific method. Astronomical problems were thereby laid open to the reason as purely mechanical ones, and the metaphysical obscurities of Aristotle and other earlier workers removed. Planets were thereafter regarded as ordinary moving bodies, and logical treatment about the nature of their orbits was made possible. The task of Sir Isaac Newton was thus made clear and definite by Galileo.

For some time after Galileo's death, few outstanding discoveries were announced, but some good progress was made. Kepler's Three Laws gave a great stimulus to theoretical speculation directed to ascertain their cause, and Galileo's use of the telescope initiated a period of steady observational work with the new instrument.

During the eighteenth century, the English instrument makers provided strong assistance to observers in England, and the quadrants and sectors of Graham, Sisson, Cary, Bird and Ramsden had no equals on the continent. Bradley's results were obtained with instruments by Graham and by Bird. The former made the Zenith Sector with which the aberration of light was discovered, and with the eight-foot quadrant Bird secured excellent Greenwich stellar observations. The art of accurate division of circular limbs into small parts was invented and perfected in England, so that observatories on the continent received their best instruments from England. Ramsden was the first to effectively substitute complete circles

Above: **Discovered in Middelburg, Netherlands, the properties of lenses led to the invention of the telescope.**

for quadrants, and Edward Troughton continued the tradition and brought instruments up to a modern standard. However, in 1804, Reichenbach's Institute was founded at Munich and this provided instruments for Germany and other continental countries equal to the best that the British made.

The principles underlying the achromatic refracting telescope were discovered by Chester More Hall about 1733, but the necessary combination of crown and flint glass and its introduction in manufacture for sale were not effected until **John Dollond (1706-1761)**, originally a Spitalfields weaver, reinvented the achromatic object glass and made instruments for the market.

Meanwhile, the reflecting telescope was a British development from start to finish. It was first invented by Newton, who made several small ones, and **John Hadley (1682-1744)** made improvements.

In a Newtonian Reflector, an image is formed by the large mirror (a paraboloid), which is obliquely reflected by a small plane mirror to the side of the tube, into an ordinary eyepiece. In the Gregorian, the observer looks straight forward, the image being thrown back by a small concave mirror through a central hole in the large paraboloid mirror, where the eyepiece is fitted. The Cassegrain is like the Gregorian, with the exception of a convex small mirror for the concave one.

James Short (1710-1768) of Edinburgh made many Gregorian telescopes which were of high quality. The Newtonian, with silvered glass mirrors instead of metallic ones—the idea of the silver-on-glass being due to a Frenchman, Jean Bernard Foucault, in the next century—has long been the favorite astronomical instrument of the amateur observer. The history of reflectors as real instruments of discovery may be said to have been started by William Herschel.

regard to fainter background stars because the Earth moved around the Sun. This had been considered before, but not in so particular a manner. Based on his study of Jupiter's satellites, he also suggested that the measurement of differences in terrestrial longitudes by means of the differences in the local times of observations of eclipses could be done.

Some of Galileo's discoveries, which have been accepted as overthrowing Ptolemy's and Aristotle's ideas for all time, met with opposition from established scholars and the Roman Catholic Church. Opposition was so strong that Galileo had to publish his expositions as mere hypotheses rather than actual systems. Conscious that opposition to his views was increasing, he visited Rome in 1611. He was given a friendly reception by the church, but fresh objections arose, and in 1615 he was secretly denounced to the Inquisition. Returning to Rome again that year, Galileo was admonished by Cardinal Bellarmine, on the pope's instructions, to abandon his opinions, and this was made the subject of a Papal Decree in 1616.

In 1623, he published his work *Il Saggiatore* (*The Assayer*) which dealt with the Copernican system in the indirect way made necessary following his admonition of 1615 and the Papal Decree of 1616. In 1632, he published his *Dialogue on the Two Chief Systems of the World, the Ptolemaic and the Copernican*. This book was really a strong advocacy of the Copernican system, and although Galileo tried to protect himself by an introduction (which today reads like a piece of irony) and by the use of dialogue, he was again examined by the Inquisition at Rome in 1633 and forced to make a retraction. He received a sentence of confinement at his country house in Arcetri, where he continued to do some scientific work.

In 1638, his eyesight, which had previously troubled him, failed completely. His blindness has been attributed to his observations of the Sun without adequate eye protection. Four years later, on 8 January 1642, he died and was buried at Santa Croce in Florence.

On 31 October 1992, Pope John Paul II formally proclaimed that the Roman Catholic Church erred in condemning Galileo for holding that the Earth was not the center of the universe. The condemnation resulted from a 'tragic mutual incomprehension,' the pope said.

Christiaan Huygens (1629-1695) of The Hague in the Netherlands was the first to construct a pendulum clock, and with the improved telescopes he constructed, he discovered Titan, the brightest satellite of Saturn, in 1655. He also found in 1656 that the appendages of Saturn, which had so puzzled Galileo, were due to a flat ring surrounding that planet. This discovery was made with a non-achromatic refractor with a 2.33-inch aperture and 23 foot focal length, the magnifying power being 100. Huygens's invention—between 1662 and 1666—of his two-lens form of eyepiece, was of great importance in the improvement of telescopes.

In his book published after his death, *Cosmotheoros*, Huygens assumed that the Sun and stars were similar bodies and that the stars were uniformly distributed in space out to infinity, each having a system of planets. By a comparison he made between the light from the Sun and that from Sirius (assumed equal to the Sun in luminosity), he estimated a distance of that star at least equal to 28,000 times the space between the Sun and the Earth, a distance which is, however, only about one-twentieth of the correct value.

Jean Dominique Cassini (1625-1712) was born Giovanni Domenico Cassini in Perinaldo, Italy, went to France in 1668 and became a naturalized citizen there in 1673. He was made the general supervisor of the Paris Observatory, where

Above: **The Italian scientist and astronomer Galileo Galilei.**

he located four new satellites for Saturn and documented the principal division in its rings, which is still known as the **Cassini Division**. He also observed that Saturn's fifth satellite, Iapetus, varied in brightness in the same period as its revolution around the planet. Although he was never completely certain, he had demonstrated that it must constantly turn the same face to its primary just as the Moon did to the Earth. Cassini was also the first to be credited with noting the **Zodiacal Light**, a nebulous cone extending from the Sun along the ecliptic, but it was actually first seen in 1683 by an English clergyman named Childrey, and a reference in Shakespeare's play *Romeo and Juliet* (Act III, scene V) suggests that it was known even earlier. The *term* Zodiacal Light, however, is due to Cassini.

His son, **Jacques Cassini (1677-1756)**, did a great deal of work on the measurement of the size of the Earth, and continued his father's investigations, particularly on the rotation of the planets.

The great English astronomer and theoretician **Sir Isaac Newton (1642-1727)** was born at Woolsthorpe near London, on Christmas Day 1642, nearly a year after Galileo's death. Even as a student, he was already at work on the projects—both practical and theoretical—that would later make him world famous. He had made many mechanical contrivances, waterclocks and model windmills. When he was only nine, he constructed a wall sundial at his home. At Cambridge University, Newton made very useful progress in mathematics and allied subjects, surprising his professors and tutors by the speed at which he absorbed what they knew. By 1666, Newton had made fundamental discoveries in mathematics—such as the **Binomial Theorem**, **Fluxions** and **Inverse Fluxions**, ie, the differential and integral calculus—by which all his, and other investigators' work in dynamics or mathematical Astronomy was made possible. He had also discovered, by experiments of passing light through a prism of glass, that white light is a composite of a range of colors, violet to red, and had also found the law of gravitation.

The discovery of the action of a prism on light formed the basis of **spectrum analysis**, to which so much chemical and astronomical knowledge is due. Unfortunately, Newton did not realize that by use of glasses of different density and powers of refraction of light, the difficulty of chromatic aberration in telescopes could be eliminated. In 1668, he invented the reflecting telescope, constructing one or two small ones himself. This type of telescope, and the related Gregorian and Cassegrainian forms, are the progenitors of the largest modern telescopes.

It was during this period—possibly while sitting under an apple tree—that Newton developed his **Law of Gravity**, which stated that every particle in the Universe attracts every other particle with a force varying inversely as the square of the distance between them, and directly as the product of the masses of the two particles. When considering the central force which is responsible for the movement of the planets in their orbits, it occurred to Newton that a force of the same nature which caused the fall of a body to the Earth, might constrain the Moon to revolve round the Earth by preventing it from flying off at a tangent, instead continuously deflecting it towards the Earth.

Newton had already seen from Kepler's First Law that for planets going round the Sun in elliptical orbits with the Sun in one of the foci of the ellipse, the attracting force towards the Sun must vary inversely as the squares of the distances. Applying this inverse square law he found, from the known distance that a body falls at the Earth's surface in a minute of time, that at the Moon's distance—which he knew to be 60 times the radius of the Earth—it would fall 16 feet. In other words, he knew that at the Earth's surface, the distance in a second is 16 feet. The space is proportional to the product of the force and the square of the time. The force varies inversely as the square of the distance from the Earth, so that the space varies as the square of the time, and inversely as the square of the distance. The distance being increased 60-fold, the space fallen though should be the same in a minute at the Moon's distance as in a second at the Earth's surface. Assuming a value for the Earth's radius, he calculated the distance through which the Moon is actually drawn, to be only 14 feet.

Philosophie Nauralis Principia Mathematics (*The Principia*), in which Newton outlined his **Three Laws of Motion**, appeared in 1687. After the publication of this book, Newton proceeded to work out some of its consequences. He perceived that the shape of a massive body must, under gravitation, be globular, and that rotation of such a globe—through the centrifugal tendency of its moving parts—would produce a flattening at the poles. He also saw that the consequent protuberance of matter at the Earth's equator would be unequally attracted by the Moon—the side nearer to it more than the other—and similarly by the Sun, but to a lesser extent owing to much greater distance.

In 1704, Newton's other great work, *Optics, or A Treatise of the Reflexions, Refractions, Inflexions, and Colors of Light*, was

SIR ISAAC NEWTON'S THREE LAWS OF MOTION

1. *Inertia*: Every body, if left to itself, free of action of other bodies, will remain at rest if it is at rest, or it will continue to move at constant velocity if it is in motion.
2. *Motion*: The rate of change of the momentum of a body measures, in direction and magnitude, the force acting on it.
3. *Reaction*: For every action there is an equal and opposite reaction.

Below, left to right: **An armillary sphere, azimuth quadrant and altitude sextant—astronomical instruments built by Tycho Brahe and accurate to within a quarter of a minute.**

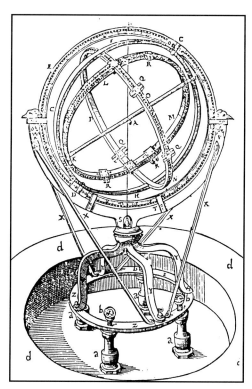 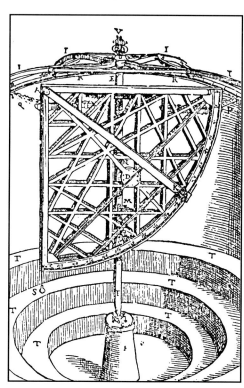 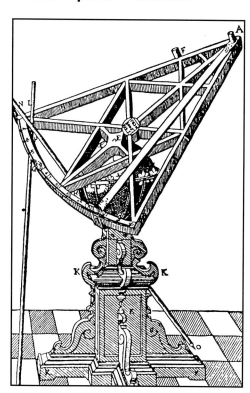

published. The first edition was published in English, unlike the first edition of the *Principia*, which was in Latin. *Optics* contains, besides his work on light, the results of many researches in chemistry and further comments on gravity. It also contains two treatises on the development of the integral calculus.

Except in Britain, the Newtonian system was accepted by very few for years after Sir Isaac Newton's death in 1727. During the late seventeenth and early eighteenth centuries, the direct followers of Newton did not develop his discoveries to any extent. This was certainly at least partly due to the difficult mathematical methods he employed, which were geometrical rather than analytical. In time, an analytical method was produced which did not require the intellect of a Newton for its use.

Sir Edmund Halley (1656-1742), a younger colleague of Newton, became Britain's Astronomer Royal in 1720. Halley, like Isaac Newton, had been something of a child prodigy. At an early age, Halley had studied Astronomy, and as a schoolboy he published a paper on the planetary orbits in 1676, which he presented to the Royal Society. Halley's chief work had been observations—the first obtained with an instrument having telescopic aid—for the compilation of a catalogue of 341 southern stars, which contained places that were an important addition to astronomical data.

Halley first met Newton in 1684, and he continued to urge him to publish his work *Principia*. Halley's own astronomical and other scientific work covered a very wide field. Using Newton's principles of the orbits, he observed and calculated the orbits of the comets of 1680 and 1682. The former is known as the one by which Newton showed that comets are controlled by the Sun's gravitation. Halley believed that it had a period of 575 years, and was perhaps identical with the bright comets seen in 44 BC and 531 and 1106 AD. However, Encke later showed that the period is probably much longer. Halley also calculated the orbits of 24 comets from which, through the resemblances of certain orbits and approximately equal intervals between appearances, he was led to suggest that the Comet of 1682—now known as **Halley's Comet**—was periodic and would return about 76 years later.

Halley also demonstrated that the acceleration of the Moon is motion, and noted periodic irregularities in the movements of Jupiter and Saturn—of which he suspected the cause. He was known to advocate the use of transits of the planet Venus over the face of the Sun for measurements determining the Sun's distance. He studied the proper motions in the sky of the stars discovered from changes in the positions of Sirius, Procyon, and Arcturus since Greek times and the association of Auroras with terrestrial magnetism. He also proposed the first general chart of the world showing the **Variation of the Compass**. He also noted the globular star clusters in Hercules and Centaurus, and considered nebulae to be composed of 'a lucid medium shining with its own proper luster,' and 'filling spaces immensely great.'

Left: This photograph of Halley's Comet was taken when it passed through Earth's atmosphere in 1910. It was named for the gifted Astronomer Royal Sir Edmund Halley, who charted among other comets, this comet's path and frequency. Halley's Comet was seen by Chinese observers as early as 240 BC, and according to Halley who saw the comet in 1682, it would return every 76 years. *Above:* As this woodcut suggests, the comet's appearance is an unforgettable event. *Right:* The zenith sector was an instrument used in the eighteenth century to measure the movement of heavenly bodies. In the process of observing the sky, astronomers discovered the aberration of light.

TRANSITS OF VENUS

Venus is seen to pass across the Sun's disc when the inclination of its orbit is sufficiently small. The occasions occur in pairs (1631 and 1639, 1761 and 1769, 1874 and 1882, 2004 and 2012). The lengths of the chords of the Sun's disc traversed by the planet, and their angular distance apart, are obtained by timing the planet's entries and exits on and off the Sun as seen from different stations on the Earth. From this data, the distance of Venus from the Earth is obtained and, the ratios of all distances in the Solar System being derivable from Kepler's Third Law, the distance of the Earth from the Sun can be calculated.

During the eighteenth century, two transits of Venus (1761 and 1769) were observed by expeditions of astronomers sent out by various governments, academies and private sources to foreign stations all over the world, while observations were also made at Greenwich, Paris, Vienna, Upsala and elsewhere in Europe. The results, as Lacaille had forecast, were not very satisfactory, owing to a difficulty in exactly determining the times of contact of the disc of Venus with the edges of the Sun's disc. The values of solar parallax at first obtained ranged from eight to 10 seconds of arc but finally, in the following century, **Johann Franz Encke (1791-1865)** deduced, from all the observations, a parallax of 8.58 minutes, corresponding to a distance of 95,370,000 miles, about two-and-a-half percent too great, but long used as the standard value.

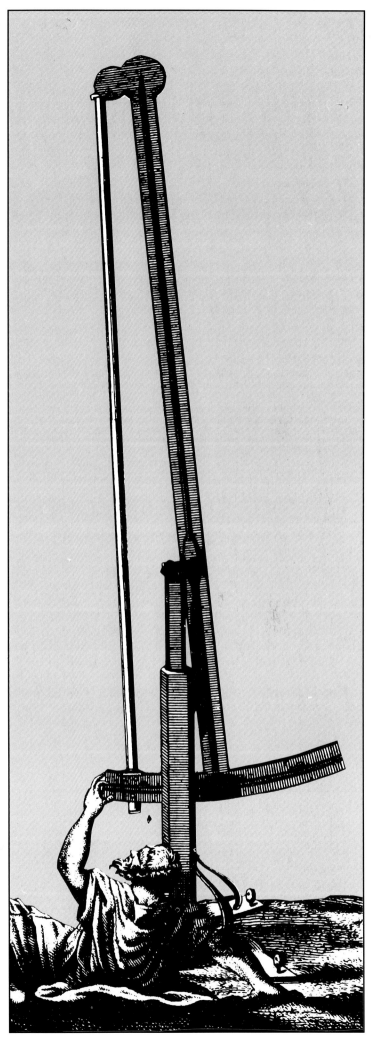

The Development of Modern Astronomy

The first major figure in the development of the astronomical science in the nineteenth century was **William Herschel (1738-1822)**. He was born Friedrich Wilhelm Herschel in Hanover, Germany, where his father was a musician in the band of the Hanoverian Guards. At the age of 14, having shown great musical talent, he joined his father's band. In 1755, when an invasion of England by the French was feared, and the Guards were sent overseas to England, and Herschel, his father and his elder brother, also a musician, moved to England. Discharged from the Guards in 1757, he spent time in London, Durham, Leeds and several other places, before going to Bath.

Herschel began to study Astronomy seriously, and in 1773 he started trying to make his own telescopes, initially constructing refracting telescopes. However, he found the long tubes intractable in use, so he took up the **reflecting telescope**, buying a two-foot long Gregorian, which he found much more convenient. He next decided to obtain a mirror suitable for a five or six-foot length telescope. It was not long before he had made the mirrors for a small telescope, and early in 1776 he finished one 5.5 feet long. Herschel's first recorded observations were made on 1 March 1774, when he viewed 'the lucid spot in Orion's sword' (the Orion nebula), and the rings of Saturn.

Herschel's earliest observational work was on the planets and the Moon. From 1774 on, he carefully observed Saturn and other planets and made drawings of Mars and the belts of Jupiter. He examined the Moon in 1776, and in 1779 he made measurements of the lengths of the shadows of lunar mountains from which their heights were derived. In 1777, he began a number of estimates of the brightness of the variable star Mira Ceti. He theorized that the changes in light of this star could best be explained by supposing rotation on an axis to bring a bright side and a dark side obscured with spots like the Sun's but larger, alternately into view.

By 1778, his interest turned to determining the annual parallax of a star using the double star method suggested by Galileo. The principle was a simple one. Depending upon an observer's shifting perspective in relation to the motion of objects at different distances from him, a star observed from opposite sides of the Earth's orbit was sometimes found to alter its situation very slightly by comparison with another star close to it in the sky, but indefinitely remote from it in space. Half of the small oscillation is called the star's **annual parallax**, and represents the minute angle under which the radius of the Earth's orbit appears at the star's actual distance. From it the distance of the star is determined. Before Herschel's time, some attempts had been made to measure **stellar parallax**, but due to the enormous distance of even the nearest stars with the resulting small displacements, and owing to the imperfection of instruments, there had been no success.

Herschel's chief purpose in his astronomical work soon settled into an investigation of the *nature* of the objects in the sky rather than their *motions*, although some of his research, such as measurements of relative positions of the components of double stars and of satellites and their primaries, were similar to the positional work of the professional astronomers. Because his solar, lunar, planetary and stellar observations were firmly directed to that end, he may be said to have been the chief originator of descriptive, and the founder of **Sidereal Astronomy**. Prior to Herschel's work, beyond measurements of position, little had been done of a systematic nature except some observation of the planets, temporary or new stars, nebulae and clusters, and variable stars.

Herschel's first preliminary review was made in 1775 with a seven-foot Newtonian telescope, and extended only down to the fourth stellar magnitude. His second review began in August 1779 with an improved seven-foot reflector of 6.2 inches aperture. He observed stars as faint as the eighth magnitude, with one of his chief objects being to compile a list of double stars apparently suitable for Galileo's parallax method.

He later submitted a paper to the Royal Society in which he wrote that 'in examining the small stars in the neighborhood of H Geminorum, I perceived one that appeared visibly larger

Above: **The Great Telescope built and used by Hevelius at Danzig (now Gdansk) was indicative of seventeenth-century technological development.** *Right:* **William Herschel is best known for his reflecting telescopes which greatly enhanced the resolution of the stars and nebulae that he observed.**

than the rest. Being struck with its uncommon magnitude, I compared it to H Geminorum and the small star in the quartile between Auriga and Gemini, and, finding it so much larger than either of them, suspected it to be a comet.'

On 17 March, he 'looked for the comet or nebulous star, and found that it is a comet, for it has changed its place.' Two days later, he ascertained that the supposed comet 'moves according to the order of the [Zodiacal] signs, and its orbit declines but little from the ecliptic.' Herschel had discovered the planet that would later be known as Uranus.

The five planets, Mercury, Venus, Mars, Jupiter and Saturn, having been known from prehistoric times, the discovery of an *all-new* sixth planet caused a considerable sensation. The fact that the new planet had been found by an amateur heightened the interest, and Herschel achieved a European reputation. The Royal Society awarded him its chief honor, the Copley Medal, in November 1781, and he was elected a fellow the following month. King George III expressed a wish to meet him, so in May 1782, he had an audience with the king, to whom he presented a drawing of the Solar System. The following July, he showed the Royal Family the planets Jupiter and Saturn and other objects through one of his telescopes.

Herschel took a special interest in Uranus, and he measured it to be between four and five times the diameter of the Earth. In 1787, he located the moons Titania and Oberon, and estimated their size to be as big as the four largest moons of Jupiter, which is incorrect. In 1797, he announced four more moons, but these later proved to be only faint stars in the planet's neighborhood at the time, as none of them were found to be identical 54 years later by William Lassell.

Herschel's second review of the sky, in which he had found Uranus, was completed in 1781. Along with it he had gathered the materials for his first catalogue of 269 double stars. He immediately began a third using the same instrument but employing a higher magnifying power of 460. Herschel carefully examined many thousands of stars as to color, singleness or duplicity, clearly defined or nebulous. When he finished the third review, he began another in 1783, using a much more powerful telescope with an 18.7-inch aperture and a 20-foot focal length. In this survey, he introduced his famous counting of stars or star gauges. In 1785, he published, in a paper to the Royal Society, the results of gauges, which showed a great deal of variation. In some, perhaps only one star at a time was counted, in others nearly 600 stars were seen, while on one occasion he made an estimate that 116,000 stars of a Milky Way region passed through his telescope field in a quarter of an hour. Generally stars were most numerous in the zone of the Milky Way and much scarcer in areas of the sky farthest from it. His final conclusion—based on nearly

Above: **Herschel's great, reflecting telescope was completed at Slough, England and put into operation in 1789.**

3000 gauges—was that the space occupied by the stars is shaped roughly like a disc or grindstone, approximately circular in plan, with its center plane coinciding with the center line of the visible Milky Way.

Herschel's first observations of double and multiple stars were made with a view to compiling a list of suitable stars for parallax measurements. However, the number he discovered and the large proportion of all stars which were accompanied by a close comparison showed, as Michell had pointed out, that most pairs must really be physically connected. In 1782, Herschel sent a catalogue to the Royal Society containing 269 star pairs, of which he had discovered 227 himself, and he made another list of 434 pairs in 1784, while his last paper to the Society in 1821 gave 145 more pairs.

By 1785, Herschel's sister **Caroline Lucretia Herschel (1750-1848)** had become well known in her right, not only as his assistant, but as an astronomer who eventually discovered eight comets and various star clusters and nebulae.

The great telescope, for which William Herschel is best remembered, took nearly four years to build and had an aperture of 48 inches and a focal length of 40 feet. Ready for use in August 1789, this instrument was not of the ordinary Newtonian form, the diagonal mirror being deleted to save the light lost in the second reflection. Herschel had experimented with a 20-foot telescope, and had found that by omitting the secondary mirror and tilting the main one, clearer, brighter images could be obtained. The telescope was thus used as a 'front-view'—later known as the Herschelian type—with eyepieces placed at the side of the top end of the tube while the observer looked down it, facing the mirror, with his back to the sky.

The fame of Herschel's work on the stars and nebulae, and on the structure of the sidereal system, tends to overshadow his solar, lunar and planetary researches. From an early date, he took an interest in solar phenomena, and he contributed several papers to the Royal Society. Herschel's observations led him to conclude that the solar globe was dark, solid and surrounded by two cloud layers, the outer one intensely hot and luminous, and the inner a sort of screen which protected the interior.

When Herschel died in 1822, the world lost one of its most important, most successful, astronomers. Herschel was the founder of **Sidereal Astronomy**. He outlined its principles, which subsequently guided its future course, and he accumulated data useful to his successors. Any *one* of his many important investigations would have been sufficient to establish a lasting reputation.

An important contemporary of Herschel's was **Johann Hieronymus Schroeter (1745-1816)**, who has been called the 'Herschel of Germany.' In 1785, Schroeter erected an observatory at Lilienthel, and in 1786, he obtained a seven-foot reflector with a six-inch aperture from the Herschel. Schroeter is credited with having begun the comparative study of the Moon's surface features, and his two books, *Selenotopographische Fragmenter*, published in 1791 and 1802, and other writings, contain many detailed observations.

Schroeter's work on the Moon, which extended from 1785 until 1813, started systematic **Selenography**—the study of the physical features of the lunar surface. The only map of any real value in existence before his time was the 7.5-inch diameter one Tobias Mayer published at Gottingen in 1775. Many previous astronomers had engaged in observations of the Moon. Herschel, for example, had studied the Moon's surface features and had tried to measure the heights of some of its mountains. Hevelius, Cassini, and Riccioli were also students of selenography, but the idea of examining lunar details persistently, with a view to detecting signs of change or activity of natural forces, was Schroeter's. He discovered the markings he named 'rills,' which appeared to be clefts in the surface about half-a-mile or so wide, a fraction of a mile deep, and up to a hundred or more miles long. He found about a dozen before 1801, and by the end of the first half of the nineteenth century more than 100 had been documented, out of the hundreds which later examination showed to exist. Schroeter even thought that he saw evidence of the existence of a lunar atmosphere in twilight extensions of the points of the lunar crescents. Schroeter was a very poor draftsman, however, and his drawings of craters and other formations have much less value than they might otherwise have had as records of the Moon's surface in the late eighteenth century.

Sir William Herschel's discoveries—especially that of Uranus in 1781—stimulated general interest, and the further development of **Descriptive Astronomy**. This branch of astronomy is more readily understood and taken up by those without special technical or mathematical training, and its progress can therefore be materially helped by amateurs without elaborate equipment. To professional astronomers, although they might often be interested in the descriptive side, there usually fell the duty of carrying on the gravitational and exact measurement.

In **Mathematical Astronomy**, **Karl Friedrich Gauss (1777-1855)** produced new and improved processes of calculation, the chief being a method of computing the orbit of a planet or comet from three or more observations of its position. This method was published in his book *Theoria Motus* in 1809. Prior to this, the prediction of the future movement of a newly discovered body was very uncertain unless the observations extended over a period of time and were numerous and accurate. Gauss's method—and his application with it of

THE DEVELOPMENT OF ASTRONOMICAL OPTICS

During the early to mid-nineteenth century, Prussian astronomer **Friedrich William Bessel (1784-1846)** made great improvements in the corrections for refraction, aberration, precession and nutation to determine actual positions, and systematized the procedure of 'reduction' of observation, not forgetting the important features of determination of errors for the instruments themselves and of the personal errors of the observer. Personal errors are errors that are peculiar to the individual observer. Maskelyne noted that one of his assistants consistently differed from him in estimating the times of transits of stars. He failed to realize that the difference was as likely to be his constant error as the assistant's, and discharged the man as incompetent. Bessel first drew attention to its nature in 1823. A psychological phenomenon, depending on the relative rates of perception by the senses, it can be measured with an instrument for the purpose, and was later minimized by a micrometer.

Most of these corrections were closely investigated in Bessel's *Fundamenta Astronomiae*, based on Bradley's observations, and he finally described his general method of reduction, and simplification of procedure, in a book entitled *Tabulae Regiomontanae* (1830).

Bessel was the first to employ the **heliometer**, an instrument constructed by **Joseph Fraunhofer (1787-1826)**, who was responsible for the improvements in lenses and mountings at the firm of von Reichenbach at Munich. The heliometer was originally intended for work on the Sun, but became the best instrument—before photography—for measuring dimensions of planets or angular distances between stars, and for stellar parallax work. The principle of this instrument, which is also called a **divided object-glass micrometer**, is the separation, by a measurable amount, of two distinct images of the same object. When a double star, for example, is under examination, the two half lenses into which the object glass is divided are shifted until the upper star in one image is brought into coincidence with the lower star in the other. Their angular distance apart then becomes known by the amount of motion employed in separating the halves of the object glass.

Bessel's improvements in methods and his construction of a great star catalogue based on Bradley's observations, were both very notable contributions. However, the most memorable of his investigations was the definite detection of the parallax of a star. For this Bessel chose the double star 61 Cygni, not because of its brightness—one obvious criterion of nearness and therefore of large parallax—but for its large proper motion of 5.2 seconds of arc per annum, which is also an indication of relative proximity. Bessel used Galileo's method, measuring with a heliometer at frequent intervals during a year, that star's distances on the sky from two neighboring stars, both faint and without sensible proper motions and thus in all probability much more distant from Earth than 61 Cygni.

At the end of 1838, he announced that the parallax was about one-third of a second of arc, corresponding to a distance of 10 light years. The remarkable accuracy of this first determination of parallax is due to the fact that the most recent value is not 10 percent different. This first parallax was quickly succeeded by two others: Alpha Centauri, nearly a second of arc or a distance of 3.25 light years, found in 1839 by **T Henderson (1798-1844)**, and Vega, about a quarter of a second, or a distance of 13 light years), by FGW Struve in 1840. The modern values for these two stars are about three-quarters of a second and one-eighth of a second, respectively. The methods used were not the same as Bessel employed for 61 Cygni, being the meridian circle and micrometer respectively, neither of which has proved so accurate as the heliometer in subsequent parallax work. Earlier observations for parallax by meridian circle or similar methods made by Piazzi, Cacciatore, Brindley, Calandrelli and others during the previous 50 years gave apparent values for Sirius, Aldebaran, Procyon, Polaris and other stars of one to several seconds of arc. But these proved to be illusory and due to periodic—sometimes seasonal—errors of the instruments employed.

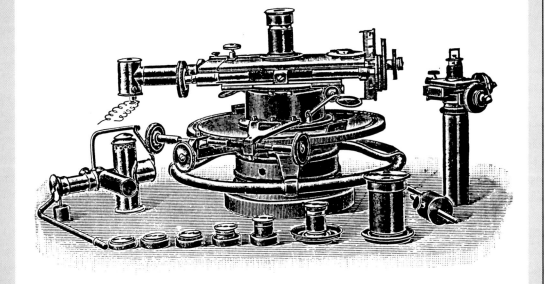

Right: **A modern micrometer with an assortment of eyepieces, oil and electric lamps. The micrometer minimizes the perceived and measurable separation of two images of the same object.**

the principle of Least Squares—soon showed its value in regard to the place of a new minor planet, which had been observed for only about six weeks before moving into that part of the sky occupied by the Sun, and had become lost afterward. Although its recorded observations covered barely three degrees in the sky, it was found nearly a year later close to the predicted position.

Another important part of Solar System science that came to be more observable in the nineteenth century were the comets. Before the introduction of the telescope in 1609, about 400 comets were recorded, all of which were bright objects visible to the naked eye. Many more were detected with optical aid, so that toward the latter half of the eighteenth century, on average nearly one comet was discovered each year. During the first half of the nineteenth century, the average became about two per year. About 30 were recorded as visible to the naked eye, the years of the appearance of nine of the brightest being 1809, 1811, 1819, 1823, 1830, 1835 (Halley's), 1843 (visible in daylight), 1845 and 1847. Two or three additional comets per year were found using a telescope. One such observer, **Jean Louis Pons (1761-1831)** detected 27 comets between 1802 and 1827. Altogether, he discovered 36.

(text continues on page 32)

THE DISCOVERY OF SPIRAL NEBULAE

One striking discovery of the period was made by the **Earl of Rosse (1800-1867)** with his 6-foot aperture reflector. This was a new class of nebula termed **Spiral Nebulae** from the shape of its spiral arms that proceeded from opposite sides of a central nucleus. The first such nebula was located in April 1845 in the constellation Canes Venatici, being number 51 of Messier's catalogue. Sir John Herschel's 18-inch reflector had shown M51 as a sort of split ring surrounding a nucleus, and it had been seen earlier as a double nebula. The M99 in the constellation Virgo was detected a year later. By 1850, 14 spiral nebulae had been recorded, the forerunners of enormous numbers to be photographed later with large reflectors.

Below: **The Spiral Nebula in the Virgo Cluster, NGC 5364.**

THE DISCOVERY OF NEPTUNE

The greatest achievement in **Gravitational Astronomy** during the first half of the nineteenth century—and probably the greatest since Newton—was the discovery of the planet Neptune in 1846. By means of the many observations of Uranus made since its discovery by William Herschel in 1781, an orbit for that planet had been calculated from which its past and future positions in the sky could be obtained. It was then found that Uranus had been seen and recorded as a fixed star by four different astronomers no less than 20 times between 1690 and 1781. However, an unexpected difficulty appeared. If these early observations were correct, then Uranus had departed from its calculated orbit, and it did not seem possible to modify the orbit so that both old and new observations would fit. As time passed, the planet got further and further away from its predicted course. For example, the discrepancies were 20 minutes in 1830, 90 minutes in 1840, and 20 minutes in 1844, using places calculated from an orbit computed by Bouvard and based on recent observations.

Inaccuracy of older observations was considered by some astronomers to be the likely explanation. Others—among them Bessel and Sir John Herschel—thought the disturbing cause might be a planet beyond the orbit of Uranus. It is of interest to note that among the first to suspect this was the Reverend TJ Hussey, Rector of Hayes, Kent, who wrote to GB Airy, then a professor at Cambridge, in 1834, suggesting that he might try to find such a planet with his 'large reflector.' Airy's reply discouraged him from this, as the motions of Uranus were, 'not yet in such a state as to give the slightest hope of making out the nature of any external action.'

There were two men who favored the outer planet idea, quite unknown to each other. These were the Englishman **John Couch Adams (1819-1892)** and the Frenchman **Urban Jean Joseph Le Verrier (1811-1877)**. Independently, they began work on the problem of determining from the observed deviations of Uranus—which never exceeded two minutes of arc, a space imperceptible to the unaided eye—the location of the disturbing planet, a problem requiring mathematical skill of the highest order. The first of the two to obtain the probable path of the outer planet, and its place in the sky at the time, was Adams. By October 1845, he had succeeded in assigning a place differing less than 2 degrees—or four times the Moon's diameter—from its actual position, and communicated his result to GB Airy, who was then the Astronomer Royal. Unfortunately, Airy was not confident as to the prediction and, under pressure from his official work at the observatory, did not have a telescopic search instituted there.

In August 1846, Le Verrier had also succeeded in obtaining an orbit and a position in the sky for the disturbing body, and when his results, which confirmed those of Adams, were brought to the attention of the Astronomer Royal, Airy requested that Professor of Astronomy at Cambridge **JC Challis (1803-1862)** undertake a search with the Northumberland 11.5-inch telescope. This was done by the laborious method of checking every star seen near the predicted place on a number of nights, to detect possible motion in the sky. In the words of S Newcomb, 'Instead of endeavoring to recognize it by its disc... his mode of proceeding was much like that of a man who, knowing that a diamond had dropped near a certain spot on the sea beach, should remove all the sand in the neighborhood to a convenient place for the purpose of sifting for it at leisure, and should thus have the diamond actually in his possession without being able to recognize it.'

In the meantime, Leverrier had written to the observatory at Berlin intimating that if a telescope were directed to a point in the ecliptic in the constellation Aquarius, at longitude 326 degrees, a planet looking like a star of about the ninth magnitude, but having a perceptible disc, could be seen. The Berlin astronomers were more fortunate than those at Cambridge in that they possessed a new star chart of the region. In fact, they may be said to have been doubly fortunate as the position of the planet was just inside the area of the chart. Consequently, shortly after commencing the search on 23 September 1846, **Johann Gottfried Galle (1812-1910)**, found the new planet within a degree of arc of the predicted place, looking like a star of the eighth magnitude, and showing a disc big enough to be measured two nights later. Galle was assisted in this discovery by **Heinrich Ludwig D'Arrest (1822-1875)**, who checked off the stars on the chart as they were located by Galle in the telescope, finally pointing out the absence from the chart of the one which was Neptune. The optical discovery, as well as the mathematical one, was thus shared by two. Meanwhile at Cambridge, Challis soon found that he had already seen the new planet. 'After four days of observing' he wrote to Airy, 'the planet was in my grasp, if only I had examined or mapped the observations.' This meant that he would have discovered it several weeks before Galle.

Within 17 days of Neptune's discovery, the English amateur astronomer and professional brewer **William Lassell (1790-1880)** had discovered its huge moon Triton with his reflector which had a two-foot aperture. It is possible that Lassell might have found the planet even before Galle. Two weeks prior to its discovery at Berlin, Lassell received a letter from WH Dawes, who had heard of Adams's results, urging him to look for a star with a disc in a certain part of the sky. Unfortunately, just about this time, Lassell sprained his ankle and could not observe. It is perhaps a pity that Dawes himself, unrivalled as an observer and using high class refractors of more than six inches aperture, did not search closer, as he might have recognized the planet by its difference from an ordinary star.

When an orbit for Neptune had been determined, it was found that, like Uranus, the planet had been seen previously as a star by several observers. In fact, on 8 May and again on 10 May 1795 **Joseph Jerome Lalande (1732-1807)** noted it as a fixed star, but suppressed the first observation as erroneous, thus losing the opportunity to discover Neptune as Herschel had found Uranus 14 years before.

Again as with the discovery of Uranus, naming of the new planet and its moon followed the mythological precedent. Because of the planet's bluish color, the name chosen was that of the roman god of the sea. Triton was a merman, the son of Greek god Poseidon—the Greek equivalent of Neptune—and the goddess Amphitrite.

Right: **The amateur astronomer William Lassell built this reflecting telescope and with it discovered Triton, Neptune's huge moon. It was operated by turning the hand crank at the base of the pier near the attendant. Lassell observed mainly faint nebulae and planetary satellites.**

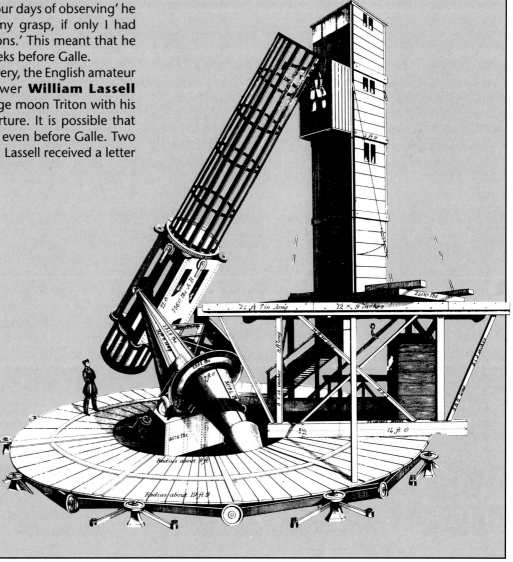

SELENOGRAPHIC DEVELOPMENTS

After Johann Hieronymus Schroeter (see page 26), the second important **selenographer**—lunar cartographer—was **Wilhelm Gottfried Lohrmann (1796-1840)**, a land surveyor in Dresden. In 1824, he published four out of 25 sections of an elaborate lunar chart on a scale of 371 inches to the Moon's diameter. His eyesight failed, however, and at his death in 1840 he left materials whereby the work was utilized and published nearly 40 years later.

The first scientifically designed map of the Moon made by a proper trigonometrical survey was issued in four parts between 1834 and 1836 by **Wilhelm Beer (1797-1850)** and **Johann Heinrich Von Madler (1794-1874)**. This work was nearly on the same scale as that produced by Lohrmann in 1824 but had greater and more accurate detail. The telescope they used was only four inches in aperture, but they were able to chart the positions of 919 formations and the heights of 1095 mountains, four of which they found to be more than 20,000 feet high.

In 1837, Beer and Madler published a descriptive volume, *Der Mond: oder allgemeine vergleichende Selenographie*, which summarized their knowledge of the Moon. They viewed it as an airless, lifeless and changeless globe (differing in this from Schroeter, Lohrmann and others), which is nearer the truth. The Moon was first successfully photographed by William Bond at Harvard University Observatory in 1850, on a small scale. Beer and Madler also made important contributions to the study of Mars. When in opposition, a planet crosses the meridian at about midnight. The Sun, Earth and Mars are in line, as viewed from North or South of the ecliptic, when Mars is in opposition. In 1830, the planet Mars came closer to the Earth than at any other time during the nineteenth century. The occasion of this favorable opposition was used by Beer and Madler to make the first systematic chart of the planet's surface features. Lines of latitude and longitude were adopted for the first time and when these findings were published in 1840, a foundation for the modern science of **areography**—Martian cartography—was laid.

The mathematician **GH Darwin (1845-1912)**, son of Charles Darwin, found by elaborate calculations, published between 1879 and 1881, that the Moon had probably originated by fission from the Earth at a time when the rotation was performed between two to four hours. The length of the day—and month—had been of this short duration before tidal friction had lengthened both the day and the month. Various objections have been advanced to this hypothesis, and it does not appear likely that it gave a correct account of the origin and evolution of the Earth-Moon system. Darwin himself wrote of his somewhat flawed hypothesis, 'It may be that science will have to reject the theory in its full extent, but it seems improbable that the ultimate verdict will be adverse to the preponderating influence of the tide on the evolution of our planet.'

In 1878, **Julius Schmidt (1825-1884)**, director of the Athens Observatory, published a lunar map based on Lohrmann's earlier results. It was 75 inches in diameter, the result of more than 30 years of observations and over 3000 drawings. On such a scale, which was twice that of Lohrmann's or Beer and Madler's maps, there could be very much more detail, about 33,000 craters of all sizes being shown on it. In 1866, he published a catalogue of 435 rills or clefts, 278 of them new.

Schmidt believed that it was possible to observe changes in the Moon, and he produced some sensation when in 1866 he announced that the small crater of Linne, formerly seen by Lohrmann and Madler over 30 years before as five or six miles in diameter, had disappeared, replaced by a white spot with a small pit only two miles wide. The reality of this change now seems doubtful, although an examination of the region in 1891 and 1892 by **William Henry Pickering (1858-1938)** using a 13-inch refractor led him to believe that lunar volcanic action was not yet completely extinct. The formation of a new crater in another area of the Moon was reported in 1878 by **HJ Klein (1844-1914)** of Cologne, but this occurrence, too, seems doubtful, particularly as the region is a complex one where changes of shadow are considerable.

Lunar photography had not sufficiently advanced to settle questions such as these, but did begin to be of great importance in selenography 12 years later in 1890, when photographs were taken with the large refractor at the Lick Observatory in California. In 1894, similar activities were undertaken at the Paris Observatory using a Coudé refractor of 23.5 inches aperture. A Coudé or 'elbowed' telescope is one in which, by use of a diagonal mirror or mirrors, the rays of light from the object glass are bent into a tube containing the eyepiece, the observer looking down it in a fixed direction.

Reproductions of the photographs taken with these two instruments were published, the Lick series as studies by **L Weinek (1848-1913)** of Prague, and those taken at Paris in an atlas with a scale of about eight feet to the Moon's diameter.

In 1864, a **Lunar Committee** was set up by the **British Association for the Advancement of Science**, which produced some valuable work by amateurs toward the preparation of a lunar map 200 inches in diameter. Owing to the death of the leading selenographer concerned, **WR Birt (1804-1881)**, this map was never completed. In the case of visual observational work, illustrations and maps were issued by **J Nasmyth (1808-1890)** and **J Carpenter (1840-1899)** in 1874, and by **E Neison (1851-1940)** in 1876, both published before the large map of Schmidt mentioned earlier. A **Selenographical Society** was founded in 1877 by Birt, Neison and others, but existed for only a few years.

In 1895, **T Gwyn Elger (1838-1897)**, a well-known English amateur, published *The Moon*, a book with a small but accurate map. Elger did not hold the somewhat extreme view of Beer and Madler that the Moon remained unchanged. His views are given in the following extracts from his book: 'The knowledge we possess, even of the larger and more prominent objects, is far too slight to justify us in maintaining that changes which on Earth we should use a strong adjective to describe have not taken place with some of them in recent years... It has been attempted to account for some of these phenomena [changes in tints of the plain areas named 'Seas'] by supposing the existence of some kind of vegetation, but, as this involves the presence of an atmosphere, the idea hardly finds favor at the present time, though perhaps the possibility

Right: **Earth's Moon.** It is the second brightest object in our sky and as such, a treasury of myth surrounds its origin and purpose. Speculation in the late-eighteenth and early-nineteenth centuries endowed the Moon with atmosphere, vegetation, seas and volcanic activity.

of plant growth in the low-lying districts, where a gaseous medium may prevail, is not so chimerical a notion as to be unworthy of consideration.'

The possibility that the Moon possessed any permanent atmosphere, even to a slight extent, seemed very unlikely. Sir John Herschel, for example, concluded, from the small difference between the measured diameter of the Moon and that derived from occultations of stars, that there was no atmosphere at the Moon's edge as much as 1/1980th of the density of the Earth's atmosphere. This conclusion was supported by a spectroscopic observation in 1865 by **Sir W Huggins (1824-1910)** of the disappearance of the spectrum of a fourth magnitude star when being occulted. This occurred as if all wavelengths of light were simultaneously equally affected—not a likely event if the light had passed tangentially through a lunar atmosphere.

In 1897, Professor Comstock of Washburn Observatory made a systematic study with a 16-inch refractor of occultations of stars, and deduced that there could be no lunar atmosphere more than 1/5000th the Earth's. The published conclusions of **G Johnstone Stoney (1826-1911)**, published in 1867, said the Moon and other comparatively small bodies cannot permanently retain an atmosphere because gas molecules, being in rapid thermal motion, would escape their gravitational attraction, and therefore preclude the formation of a permanent atmosphere.

On the other hand, there is one circumstance, pointed out by **AC Ranyard (1845-1894)**, which would appear to favor the existence of some atmosphere, however tenuous: The unworn appearance of lunar features suggests protection from meteoric bombardment. It is interesting to note the different view taken many years later by CP Olivier, the American authority on meteors, as to the result of meteor falls on the Moon. He remarked, 'The general effect of all this, be it great or small in amount, must be somewhat to smooth the lunar surface. Had it not acted at all, that surface would be even rougher than it is now.'

It should be borne in mind that a lunar atmosphere of small total quantity, and very great tenuity, would have a density, at heights above the Moon's surface similar to those where terrestrial meteors are consumed, not unlike the density of the Earth's atmosphere there, this being due to the much smaller gravitational attraction of the Moon on any atmosphere it might have. In 1972, the American Apollo 17 astronauts found slight traces of hydrogen and helium being exuded by the lunar surface. These traces were so faint that they had remained undetected by the five previous Apollo landing teams.

Experiments on heat radiating from the Moon were first carried out by Melloni as far back as 1846, when sensible effects were noticed. Between 1869 and 1872, the Earl of Rosse measured lunar radiation by means of his three-foot aperture reflector, and found that the results indicated an actual heating of the lunar surface, which he estimated to slightly under 100 degrees Centigrade.

In 1887 however, **Samuel Pierpont Langley (1834-1906)** deduced a very much lower temperature from his experiments, in fact that of freezing water. FW Very's results, published in 1898, were that the Moon's surface under vertical solar radiation, is hotter than boiling water, while its temperature sinks during the 14-day Lunar night to about that of liquid air. These results by Rosse and Very have been shown by more refined modern methods to be generally accurate.

BODE'S LAW

	Distance	
	Calculated	Real
Mercury	0+4=4	3.9
Venus	3+4=7	7.2
Earth	6+4=10	10.0*
Mars	12+4=16	15.2
**	24+4=28	
Jupiter	48+4=52	52.0
Saturn	96+4=100	95.5
Uranus	192+4=196	192.2
Neptune	---	300.1
Pluto	384+4=388	394.6

In 1800, Bode's countryman Johann Hieronymus Schroeter organized what he called the **Celestial Police**, an association of astronomers dedicated to finding the planet whose existence had been postulated by Bode (and by Titius of Wittenberg before him). Ironically, the first, and largest, asteroid, Ceres, was not discovered by Bode or by a member of Schroeter's Celestial Police, but was discovered on New Year's Day in 1801 by Giuseppe Piazzi, director of the observatory at Palermo, Sicily. Though Piazzi later joined the Celestial Police, the most successful member was **Heinrich Olbers (1748-1840)**, a German amateur astronomer who was able to 'recover,' or rediscover, Ceres in 1802. Olbers then went on to discover the asteroids Pallas and Vesta in 1802 and 1807 respectively. It would be over 30 years, however, before the next asteroid would be discovered. Olbers later advanced the hypothesis that the minor planets were fragments of an exploded large planet.

In 1830, another German amateur astronomer, **Karl Ludwig Hencke (1793-1866)**, went on a search for further asteroids which finally bore fruit in 1845 with the discovery of Astraea. Two years later Hencke discovered a sixth asteroid, Hebe. By the 1840s photography was brought into play as a tool and suddenly the search for new asteroids took on a whole new flavor. Both Iris and Flora were discovered in 1847, the same year that Hencke found Hebe, and after that several were found each year. Within ten years, 48 asteroids had been discovered, and by 1899 there were 451 known asteroids. By 1930, the year that Clyde Tombaugh discovered Pluto, there were more than 1000 known asteroids. After World War II, the International Astronomical Union set up a cooperative program of asteroid research. By 1980, there were more than 2000 known asteroids, and by the end of 1990 the number exceeded 3500.

*The distance between the Earth and the Sun was arbitrarily fixed at 10.0.
**The reason for positing a gap between Mars and Jupiter is that the asteroids fill this gap.

In some ways, the most remarkable comets of the period were not the large, naked-eye ones. One comet, seen by Pons in 1818, was named for Encke because of the great amount of mathematical work done by Encke on its orbit. It was periodic and had the shortest period known at that time—3.3 years. Encke found that, after making all the necessary allowances for perturbations by the planets, there remained an outstanding continuous but variable shortening of the period, which averages about one and one-sixth days per revolution, that was attributed to the effect of resistance of some kind to the comet's movement. If a resisting medium is responsible, the fact that the other short-period comets do not show a shortening of period may be explained by the medium not extending beyond the orbit of Mercury, Encke's being the only short-period one which passed inside that orbit when close to the Sun. On the other hand, the apparent absence of effect on the motion of parabolic, or very long period, comets which pass even closer to the Sun, must be noted in this connection.

A comet called Biela's Comet, with a period of 6.75 years, was observed in 1772, 1805, 1826, 1832, without anything remarkable, but in 1846 it became pear-shaped and actually divided into two parts which traveled together for over 3 months separated by a distance of 160,000 miles, each part developing a nucleus and a tail. When observed in 1852, the separation was nearly 10 times as much and neither part has been seen again.

Another remarkable cometary occurrence was the passage of the Earth through the tail of the comet of 1819 with no noticeable consequences. Olbers had suggested in 1812 that the tails of comets are rapid outflows of highly rarefied matter, most of which becomes permanently detached from the comet's head or nucleus, their formation being due to some form of solar electrical repulsion.

The 1835 appearance of Halley's comet was—like that of 1910—best seen in the Southern hemisphere, where it was well observed by Sir John Herschel at the Cape of Good Hope. The 1843 comet, visible in daylight, seems to have been the most brilliant of the comets seen during this period.

The study of shooting stars, or **Meteoric Astronomy**, may thus be said to have become a definite department of the science from the first half of the nineteenth century. Until the beginning of the nineteenth century the most commonly held view of the nature of meteors or shooting stars was that they were really meteorological phenomena caused by the ignition of vapors in the lower atmosphere. Halley apparently held this belief that they came from outer space, but that view was certainly not a general one. The physicist **EFF Chladni (1756-1827)**, however, in 1794 expressed the opinion that meteors are small bodies circulating through space, becoming visible through incandescence on entering the Earth's atmosphere. Acting on his suggestion, in 1798 two students of Gottingen University, Brandes and Benzenberg, observed a number of meteors simultaneously from separate observation points, and, using the base line thus provided, they calculated that meteors appeared in the atmosphere at great heights moving with speeds like those of the planets in their orbits. Some astronomers, including Laplace in 1802, maintained that meteors were ejected from craters on the Moon.

No progress was made in elucidating the subject for the first 30 years or so of the century. Such ideas as existed were, however, clarified by investigations following the extraordinary shower of meteors on 12 November 1833, when nearly a quarter of a million were estimated to have been seen by

Above: This iron meteorite weighing approximately 40 pounds was found in Antarctica by a team of researchers. *Right:* The Comet Kohoutek streaked across the sky in January 1974.

observers in America during a nine-hour display. All of these shooting stars appeared to come from the same part of the sky—the constellation Leo—with their paths diverging from one radiant point. As a result, it was soon deduced that these meteors moved in parallel paths round the Sun, apparently divergent in the sky because of perspective.

Actual falls of meteorites to the Earth's surface also had occurred at L'Aigle in France in 1803, consisting of between 2000 and 3000 small, stony meteorites, and at Stannem in Austria in 1808, when about 200 stones fell. Among many falls of single meteorites at the time were one in 1830 at Taunton and one in 1835 at Aldsworth, both in England. These drew attention to another aspect of the study, in the examination and analysis of the constitution of such bodies.

As the time approached for the next brilliant display, judged by the interval observed between previous great ones (1799 to 1833), **Professor HA Newton (1830-1896)** of Yale University in Connecticut, examined all the past records he could find and noted that great displays had been seen between 902 and 1932 in the following years: 934, 1002, 1101, 1202, 1366, 1533, 1602, 1698 and 1799. He predicted another great display for the night of 13-14 November 1866. This prediction was fulfilled, although the shower was not so great as in 1833.

It had been found that great showers every 33 or 34 years could be explained by five different periods, combined with varying lengths of extension of the swarm along the orbit, the shortest period being 354.5 days, the longest 33.25 years. It required great mathematical powers to solve the problem, as the gravitational disturbances of the various planets had to be computed. By a very arduous process of calculation, John Couch Adams, co-discoverer with Le Verrier of Neptune, ascertained that all the observed phenomena would be accounted for satisfactorily only by the longest period, 33.25 years. When ancient records were examined, it was found that the date of return of the shower had gradually changed, with an alteration of the orbit, the meteors at each return crossing a point in the Earth's path about half a degree further on in the direction of the Earth's movement. If the orbit was the largest of the five, this shift would be due to the action of Uranus, Saturn, Jupiter and the Earth, while if it was one of the smaller four the planets responsible would be Jupiter, the Earth and Venus. Adams found that with the smaller ellipses, it was impossible to obtain a displacement of half the amount observed, but that with the largest, Jupiter would account for two-thirds of the change and Saturn for most of the remainder, with a small part due to Uranus.

Sir John Herschel pointed out in *Outlines of Astronomy* that if the orbits were of the short period of nearly a year, the meteors must have been so much encountered by the Earth as to have been completely scattered into orbits of all degrees of inclination and eccentricity. The demonstration was therefore clear that the largest of the five orbits was the correct one.

In 1866 **Giovanni Schiaparelli (1835-1910)** computed the orbit of the August meteors and found it was the same as that of a comet observed in 1862, and in 1867 he found that the revised orbit of the Leonid meteors as calculated by Le Verrier was about identical to that of a comet seen in 1866. In 1948, the French astronomer Guigay showed that, along with the comet of 1862 and other comets seen in 1825, 1826, 1877 and 1932, this Perseid shower probably forms the remains of a disrupted large comet.

Biela's Comet has also been connected with a periodic shower of meteors with its radiant in Andromeda. There are several other cases of similar association between periodic comets and meteors, the comets concerned being Halley's, Encke's, Winnecke's and that known as Giacobini-Zinner.

A von Humboldt (1769-1859) at Cumana in Venezuela, who had seen nearly as rich a display in the year 1799, had noted that the meteors also diverged from a radiant, did not attempt anything by way of an explanatory theory. The first to suggest that these 'Leonid' meteors might be particles in a swarm moving around the Sun was Olmstead of Yale University. He proposed for them a period of about six months in an orbit nearly intersecting that of the Earth at the point occupied by it on 12 November. The next to propose an explanation was Olbers. In 1837, he suggested a period of 34 years basing this on Humboldt's 1799 observation.

THE ASTEROIDS

Although in planetary Astronomy the chief event of the first half of the century was the discovery of another member of the Solar System, Neptune, there were other discoveries of planets of a different kind. These were perhaps of more significance with respect to theories of the origin and development of the planetary system.

On the first evening of the nineteenth century, 1 January 1801, **Giuseppe Piazzi (1746-1826)**, a professor of mathematics and Astronomy at Palermo, Sicily, noted the position of an eighth magnitude star in the constellation Taurus, a part of the sky which he was observing in connection with an error he had found in a star catalogue. It was his practice to focus on a set of 50 stars for four successive nights, and on the two following nights, he found that the star had changed its position. He continued to observe it whenever possible until 11 February when he unfortunately fell ill.

Gauss indicated that this new body belonged to the space between the orbits of Mars and Jupiter, which had been noted as abnormally large by astronomers since Kepler's day. That a body might be found in this gap had appeared probable from an empirical relation between the sizes of the planets' orbits first pointed out by **JD Titius (?-1796)**. This Wittenberg professor had appended a note to a translation he had made of a scientific work, published in 1772, which drew attention to the existence of a series of numbers that expressed, with rough accuracy, the distances from the Sun of the six then known planets. The numbers are obtained by adding 4 to 0, 3, 6, 12, 24, 48 and 96, the results are 4, 7, 10, 16, 28, 52 and 100, fairly representing the actual distances except that 52 and 100 have to be taken for Jupiter and Saturn, with no planet at 28. The discovery of the field of minor planets in the 342 million mile interval between the orbits of Mars and Jupiter that we know as the Asteroid Belt dates to the theoretical work of German astronomer **Johann Elert Bode** of the Berlin Observatory.

In 1772, Bode authored **Bode's Law**, which took into account the regular intervals between the known planets and postulated that a planet should, by his law, exist between Mars and Jupiter. Little did Bode realize that this interval was not filled by a single planet, but rather by thousands of planetoids, or as we know them today, **asteroids**. Bode noted the absence of a planet to correspond with the theoretical planet at the number 28, and he organized a combination with five other astronomers for the purpose of trying to find what they thought might be a new planet.

While their name translates as implying a star-like character, asteroids are more accurately described as planetoids or minor planets. Literally they are fragments of rock that may have their origin in the cataclysmic destruction of one or several terrestrial planets, or they may be debris left over from the origin of the Solar System itself. The largest asteroid, Ceres, is 622 miles in diameter, and there are only seven known asteroids with diameters greater than 200 miles.

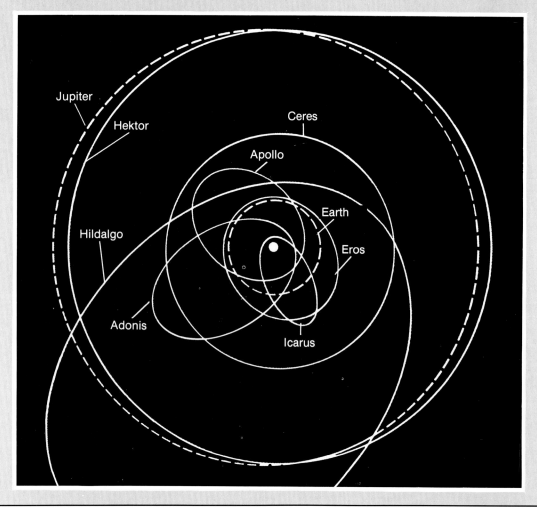

Left: Asteroids are minor planets, most of which orbit the Sun between Mars and Jupiter in a region called the Asteroid Belt. This diagram represents the aphelion and perihelion of the elliptic orbits of a few notable asteroids. The Sun is at the center with Earth's orbit and Jupiter's orbit indicated by broken lines. Ceres is the largest asteroid with a diameter of 621.86 miles.

THE FIRST TEN ASTERIODS DISCOVERED

Rank	Name	Absolute Magnitude	Discoverer	Date of Discovery	Closest Distance to Sun (AU)	(Miles)	(Km)	Farthest Distance from Sun (AU)	(Miles)	(Km)	Diameter (Miles)	(Km)	Sidereal Period in Earth Years
1	Ceres	4.5	Piazzi	1 Jan 1801	2.55	237,150,000	381,811,500	2.94	273,420,000	440,206,200	621.86	1003	4.60
2	Pallas	5.0	Olbers	23 Mar 1802	2.11	196,230,000	315,920,300	3.42	318,060,000	512,076,600	376.96	608	4.61
3	Juno	6.5	Harding	1 Sep 1804	1.98	184,140,000	296,465,400	3.35	311,550,000	501,595,500	155.00	250	4.36
4	Vesta	4.3	Olbers	29 Mar 1807	2.15	199,950,000	321,919,500	2.57	239,010,000	384,806,100	333.56	538	3.63
5	Astraea	8.1	Hencke	8 Dec 1845	2.10	195,300,000	314,433,000	3.06	284,580,000	458,173,800	62.54	117	4.14
6	Hebe	7.0	Hencke	1 Jul 1847	1.93	179,490,000	288,978,900	2.92	271,560,000	437,211,600	120.90	195	3.78
7	Iris	6.8	Hind	13 Aug 1847	1.84	171,120,000	275,503,200	2.94	273,420,000	440,206,200	129.58	209	3.68
8	Flora	7.7	Hind	18 Oct 1847	1.86	172,980,000	278,497,800	2.55	237,150,000	381,811,500	93.62	151	3.27
9	Metus	7.8	Graham	25 Apr 1848	2.09	194,370,000	312,935,700	2.68	249,240,000	401,276,400	93.62	151	3.69
10	Hygiea	6.5	DeGasparis	12 Apr 1849	2.84	264,120,000	425,233,200	3.46	321,780,000	518,065,800	279.00	450	5.60

OTHER ASTEROIDS

Rank	Name	Absolute Magnitude	Discoverer	Date of Discovery	Rank	Name	Absolute Magnitude	Discoverer	Date of Discovery
11	Parthenope	7.8	DeGasparis	11 May 1850	46	Hestia	12.5	Pogson	16 Aug 1857
12	Victoria	8.4	Hind	13 Sep 1850	47	Aglaia	12.9	Luther	15 Sep 1857
13	Egeria	8.1	DeGasparis	12 Nov 1850	48	Doris	12.1	Goldschmidt	19 Sep 1857
14	Irene	7.5	Hind	19 May 1851	49	Pales	12.7	Goldschmidt	19 Sep 1857
15	Eunomia	6.4	DeGasparis	29 Jul 1851	50	Virginia	13.6	Ferguson	4 Oct 1857
16	Psyche	6.9	DeGasparis	17 Mar 1852	51	Nemausa	11.2	Laurent	22 Jan 1858
17	Thetis	9.1	Luther	17 Apr 1852	52	Europa	11.7	Goldschmidt	4 Feb 1858
18	Melpomene	7.7	Hind	24 June 1852	53	Calypso	13.1	Luther	4 Apr 1858
19	Fortuna	8.4	Hind	22 Aug 1852	54	Alexandra	12.2	Goldschmidt	11 Apr 1858
20	Massilia	7.7	DeGasparis	19 Sep 1852	55	Pandora	12.1	Scarle	10 Sep 1858
21	Lutetia	8.6	Goldschmidt	15 Nov 1852	56	Melete	9.5	Goldschmidt	9 Sep 1857
22	Calliope	7.3	Hind	16 Nov 1852	57	Mnemosyne	8.4	Luther	22 Sep 1859
23	Thalia	8.2	Hind	15 Dec 1852	58	Concordia	9.9	Luther	24 Mar 1860
24	Themis	8.3	DeGasparis	5 Apr 1853	59	Elpis	8.8	Charcornac	13 Sep 1860
25	Phocea	9.3	Charcornac	6 Apr 1853	60	Echo	10.0	Ferguson	15 Sep 1860
26	Proserpina	8.8	Luther	5 May 1853	61	Danae	8.9	Goldschmidt	9 Sep 1860
27	Euterpe	8.4	Hind	8 Nov 1853	62	Erato	9.8	Lesser	14 Sep 1860
28	Bellona	8.2	Luther	1 Mar 1854	63	Ausonia	9.0	DeGasparis	11 Feb 1861
29	Amphitrite	7.1	Marth/Pogson	1 Mar 1854	64	Angelina	8.8	Tempel	6 Mar 1861
30	Urania	8.8	Hind	22 Jul 1854	65	Cybele	8.0	Tempel	10 Mar 1861
31	Euphrosyne	7.3	Ferguson	1 Sep 1854	66	Maia	10.5	Tuttle	10 Apr 1861
32	Pomona	8.8	Goldschmidt	26 Oct 1854	67	Asia	9.7	Pogson	18 Apr 1861
33	Polyhymnia	9.9	Charcornac	28 Oct 1854	68	Leto	8.2	Luther	19 Apr 1861
34	Circe	9.6	Charcornac	6 Apr 1855	69	Hesperia	8.2	Sciaparelli	29 Apr 1861
35	Leukothea	9.7	Luther	19 Apr 1855	70	Panopea	8.9	Goldschmidt	5 May 1861
36	Atalanta	9.8	Goldschmidt	5 Oct 1855	71	Niobe	8.3	Luther	13 Aug 1861
37	Fides	8.4	Luther	5 Oct 1855	72	Feronia	10.1	Peters	29 Jan 1862
38	Leda	9.7	Charcornac	12 Jan 1856	73	Clytie	10.3	Tuttle	7 Apr 1862
39	Laetitia	7.4	Charcornac	8 Feb 1856	74	Galatea	10.0	Tempel	29 Aug 1862
40	Harmonia	8.3	Goldschmidt	1 Mar 1856	75	Eurdice	10.0	Peters	22 Sep 1862
41	Daphne	8.2	Goldschmidt	22 May 1856	76	Freia	9.1	D'Arrest	14 Nov 1862
42	Isis	8.8	Pogson	23 May 1856	77	Frigga	9.7	Peters	12 Nov 1862
43	Ariadne	9.2	Pogson	15 Apr 1857	78	Diana	9.2	Luther	15 Mar 1863
44	Nysa	7.8	Goldschmidt	27 May 1857	79	Eurynome	9.2	Watson	14 Sep 1863
45	Eugenia	8.3	Goldschmidt	28 Jun 1857	80	Sappho	9.2	Pogson	3 May 1864

The Science of Astrophysics

With the exception of the inventions of the telescope and of photography, no discovery has given so great an impetus to Astronomy as the instrument used in spectrum analysis—the **spectroscope**. Spectrum analysis is analysis of light from a luminous body by means of a prism, or set of prisms, or by a diffraction grating. If the light of the Sun is admitted through a narrow slit, an inverted image of the slit can be formed on a screen opposite by means of a lens of appropriate focal length. When a prism of glass is interposed, the image is deflected to one side. Newton had showed that the images of the different colors making up white light are deflected by varying amounts—the red least, the violet most. A continuous band of light—or spectrum—is thus formed which shows all the colors of the rainbow: red, orange, yellow, green, blue, indigo and violet.

Below: **A multi-prism spectroscope.** *Opposite:* **The Great Refractor is the Harvard College Observatory's 15-inch telescope.**

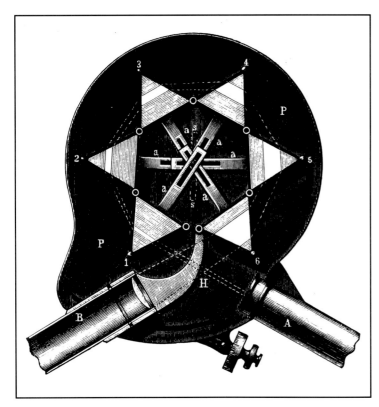

In 1802, WH Wollaston had observed seven dark lines across this spectrum, which he took to be the edges of the primary colors. In 1814-1815, Fraunhofer studied the spectrum much more thoroughly and found about 600 lines, mapping the position of 324. Shortly afterward, he examined the spectra of the Moon, Venus, Mars and the fixed stars Sirius, Castor, Pollux, Capella, Betelgeuse and Procyon. For the Moon and planets he found spectra very similar to that of sunlight, but for the stars there were differences. When artificial lights were similarly analyzed, he found that incandescent solids or liquids, or dense gases, had a continuous spectrum, while incandescent gases of low density had bright lines. He also found that dark bands or dark lines could be produced artificially in continuous spectra by passing the light through various substances or gases.

Earlier experiments by Melvil in 1752 had shown that the light from flames tinged with metals or salts gave characteristic bright lines, and in 1823 Sir John Herschel suggested these lines might be used as a test for the presence of the substances concerned. Various explanations for the phenomena were attempted, but the first to publish a satisfactory one was **GA Kirchhoff (1824-1887)** in 1859, although a correct one had been adopted by G C Stokes of Cambridge, who had treated it as more or less an academic question. Kirchhoff repeated Fraunhofer's experiments and found that if a spirit lamp with common salt in the flame were placed in the path of the Sun's light, a dark line already in the spectrum (called 'D' by Fraunhofer) was intensified.

Kirchhoff also found that if he used a limelight instead of sunlight and passed it through the flame with salt, a dark D line appeared in the otherwise perfectly continuous spectrum. However, it appears that **Jean Bernard Foucault (1819-1868)** had obtained similar results in 1849, using a voltaic arc. This was unknown to Kirchhoff.

Kirchhoff explained his findings as due to the absorption by sodium vapor in the flame—sodium being a constituent of salt—absorbing the same light as it emits at high temperature, and that vapor itself then giving a bright line in the place in the spectrum at which the D line appears. He said this proved that sodium vapor existed in the Sun's atmosphere. His subsequent experiments showed that the sun actually has iron, magnesium, calcium, and chromium, along with copper, zinc, barium and nickel in smaller quantities. There is no reason to suppose that any of the known natural elements are

not present. This was the dawn of the science of **Astrophysics**.

The accumulation of facts regarding the constitution of the atmosphere of the Sun, stars and planets, and of the nebulae was thus initiated. **Christian Johann Doppler (1803-1853)**, an Austrian professor of mathematics in Prague, pointed out in 1842 that if a luminous body is approaching an observer, its waves of light were crowded together—with a slight shift in the direction of the violet end of the spectrum—and the opposite effect occurs in a receding body. From this principle, by comparing the spectrum of a heavenly body alongside that of the spectrum of iron or another element obtained in the laboratory, a displacement of spectral lines can be measured, and the velocity of approach or recession of the body ascertained. If the displacement is towards the violet, the motion is towards the observer, and if the displacement is toward the red, it is one of recession. Correction, of course, must be made for the motion of approach or recession of the Earth with respect to the body, which varies with the Earth's motion in its orbit around the Sun. Therefore, not only was there an instrument for determining the elements present in celestial bodies, but it functioned as a powerful adjunct to the study of the movements of those bodies using their proper motions in the sky, from which excellent results could later be obtained.

It is believed that photographic pioneer Louis Daguerre, or one of the astronomers at the Paris Observatory, took a daguerreotype of an eclipse of the Sun in 1839 or 1840. Draper took a successful photograph of the Moon in 1840 with 20 minutes exposure, and, in 1843, one of the solar spectrum. In 1845, Fizeau and Foucault obtained one of the Sun showing groups of spots.

In 1850, a noteworthy achievement was recorded at Harvard College Observatory, where daguerreotype photographs were secured of Vega and Castor by means of the 15-inch refractor. One or two minutes exposure was required with such primitive photography equipment, and due to the unsteady movement of the clockwork driven instrument, second magnitude stars were beyond its power. Within seven years, improvement in the steadier movement of the telescope and use of the collodion process enabled photographs to be taken of all stars seen with the naked eye.

Before the end of the century, photography became the equal of eye observation in all but lunar and planetary Astronomy, and superior in the study of nebulae, comets, stellar distribution and measurements of stellar positions. Substitution of silver-on-glass mirrors for metallic specula, an invention of the German optician **CA Steinheil (1832-1893)** and, independently, of the French physicist Foucault about 1856, increased the efficiency of reflectors as light-gathering instruments by more than one-third. The size of the refracting telescope increased rapidly with improvement in the manufacture of optical glass. The maximum of 15 inches aperture in 1850 evolved to 40 inches by the end of the century, in the case of the Yerkes Telescope at the University of Chicago. An idea of the growth in dimensions may be found in the following examples: Newall, 25-inch (1869); Vienna, 27-inch (1880); Pulkowa, 30-inch (1885); Lick, 36-inch (1888); and Yerkes, 40-inch (1897).

The increased size of the *reflector* was also noteworthy. Leaving out of account the six-foot Rosse metallic speculum, the largest reflector built by 1900 was a 60-inch aperture silver-on-glass instrument, made in 1888 by amateur enthusiast Dr **AA Common (1841-1902)** of London, which did valuable service in the photography of nebulae. Other reflecting telescopes of particular note were: Melbourne, 48-inch (1870), a metallic speculum, (Cassegrain); Paris, 47-inch (1875); Lick (Crossley), 36-inch (1879); Cambridge, 36-inch (1890); Toulouse, 33-inch (1887); Marseilles, 32-inch (1873); and Greenwich, 30-inch (1897).

The regular growth in the maximum size of the *refractor* was due to the progress in ability to make optical glass in large discs. In the case of the reflectors, there was no such limiting factor operating, as the discs for them were not required to be optically perfect for the purpose of *transmitting* light, as with refractors.

A new development was that of the photographic refractor with its object glass corrected so as to bring the blue and violet

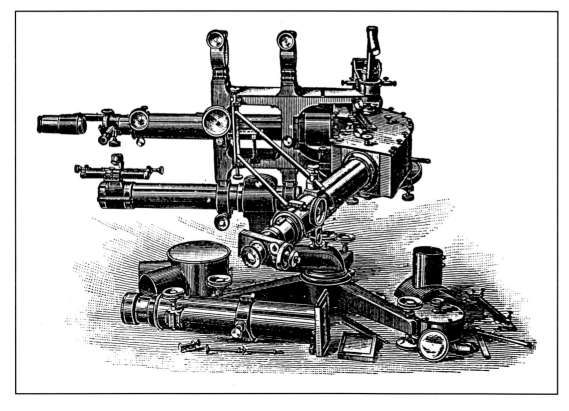

Left: **This universal spectroscope by Brashear is arranged for viewing, although below it are the tube and plateholder, which are installed when spectra are to be photographed. Photography of nebulae, comets, stellar positions and stellar distribution along with improved telescopes enabled late nineteenth-century astronomers to collect information on the velocity and composition of what was beyond the range of human sight.**

rays of strongest actinic effect, rather than those used visually, to be as nearly as possible the same focus. These instruments, with their great focal length and large scale image, were particularly useful in stellar parallax work, for which the ordinary type refractor was also used with a color screen to limit the range of wavelengths acting on the photographic plate. Wide angle photographic camera objectives of three or four lenses, that gave good definition over a wide field of the sky, also made their appearance, starting with ordinary portrait lenses and evolving into an instrument such as the Bruce 24-inch, made in 1893, which had a relatively short focal length. It was later taken to the Harvard College Observatory Station at Bloemfontein, South Africa.

Up to this time, instruments for exact measurement were usually confined to two types: those set up in the north plane of the meridian so as to intercept stars at their rising points, and those mounted equatorially with clock drive for objects in their diurnal motion. Several notable additions were made to these two chief methods of using the telescope, such as the altazimuth, which can measure positions of the meridian in any part of the sky, and a prime vertical instrument set up in the plane at right angles to the meridian.

In **Gravitational Astronomy**, there were no startling results such as the discovery of Neptune, but some important research was undertaken successfully during the late nineteenth century. John Couch Adams discovered that the amount should have been five seconds or six seconds instead of 10 seconds. An explanation for this discrepancy was first proposed by **C Delaunay (1816-1872)**, who determined it was due to the lengthening of the day by about one-thousandth of a second per century. An attempt to calculate the amount of lengthening caused by tidal friction was made in 1853 by W Ferrel. It was not until more than a half century had passed, after calculations had been made by GI Taylor and by H Jeffreys for the regions of the oceans where tides are abnormally strong, that Delaunay's suggestion became the generally accepted explanation for at least part of the phenomenon first investigated with success by Laplace.

It was found that, on the average, the dissipation of the rotational energy of the Earth is equal to about 1500 million horsepower, two-thirds of this being due to the tides of the Bering Sea, where currents are strong and the water shallow. The total energy necessary for the lengthening mentioned would be about 2.1 billion horsepower.

Another mathematical result of outstanding interest was reached by **James Clerk Maxwell (1831-1879)** who, in 1857, showed that the rings of Saturn could not be circulating solid or liquid appendages but could maintain their form and stability only if they were composed of an enormous number of small bodies revolving in independent orbits around Saturn. A similar theory had been suggested by GP de Roberval in the seventeenth century, Jacques Cassini and Wright of Durham in 1750, but their suggestions were merely speculative. Laplace's researches had shown only that the rings could not be a uniform solid body unless their weight was unsymmetrically disposed, but he had left it at that.

In 1867, **Daniel Kirkwood (1815-1895)** had suggested that the Cassini division between the outer and inner ring could be explained by the perturbations of some of the satellites. In 1888, it was shown by Seeliger that only a ring composed of particles would show the observed surface brightness of the rings under varying angles of illumination by the Sun. **JE Keeler (1857-1900)** demonstrated by spectroscopic measurements of the radial velocities of its different parts that the constitution was undoubtedly as indicated by the mathematical researches of Maxwell. The inner edge of the rings appeared to move at 12.5 miles per second in the line of sight, while the outer edge had an apparent speed of 10 miles per second. These velocities are in accordance with what would be the case for separate satellites at the respective distances from Saturn. For a solid or liquid ring the speed of the outer edge would be *greater* and not less, than that of the inner edge.

In 1850, Roche of Montpelier proved that no liquid satellite could exist as a coherent body inside about two to four times a planet's radius, if the planet and its satellite were equally dense. Mimas, the nearest of Saturn's satellites, was at three to 11 times its radius, while the outer edge of the rings was only 2.3 radii out. This suggested that the rings might be satellite-forming material in a position where no satellite was possible. However, in 1947, Jeffreys found that a *solid* satellite of rock moving near Jupiter's surface will not break up unless it is greater than about 250 miles in diameter. For a rocky satellite near to the Earth, the limit is about 130 miles.

During the late nineteenth century, the distance to the Sun was measured by various methods. Transits of Venus in 1874 and 1882 were observed at widely scattered stations on the Earth, but the resulting values obtained were disappointingly diverse. The difficulty of consistently and accurately judging times of contact of the planet's edges with the limbs of the Sun prevented truly reliable results, so the variety of the distances derived was considerable. For the 1874 transit, such differences ranged from about 90 to 93.5 million miles, and for that of 1882, from 91.75 to 92.5 million.

Meanwhile, other methods were being exploited. In one of these the planet Mars and certain minor planets played a part. The observation of Mars at opposition was employed several times with success. One of these occasions was in 1877 when **Sir David Gill (1843-1914)** observed it on Ascension Island.

The method employed was originated by Flamsteed. It was tried by **WC Bond (1789-1859)** in America and again suggested by Airy in 1857. In order to find the distance of a planet from the Earth, it was necessary to observe it from two different terrestrial stations separated by a known space. This could normally be done by observing from two positions as much apart in latitude as possible. Airy's suggestion was that if observations are taken, at an interval of several hours, from the same station, the Earth's rotation would provide a base line of a known size. The apparent shift of the planet in the sky would be found by measuring its angular distances on the sky from fixed stars nearby. This method has been also used with minor planets, which, although usually further from the Earth than Mars, are easier to measure as they have no perceptible disc.

In 1872, Galle, the telescopic discoverer of Neptune, was the first to suggest the use of the minor planets in this connection. By using the distance from the Earth to Mars, and of the particular minor planets utilized, the Sun's distance from the Earth can be computed. This was because Kepler's Laws governing the motions of the planets enabled the relative distances from the Sun of the planets to be inferred from the periodic times in which their orbits were described. The Solar System could, as it were, be accurately drawn to plan, and the scale of the plan determined when any one distance in it has been measured. By Kepler's Third Law, the cubes of the distances from the Sun are proportional to the squares of the periodic times of revolution. The planet Mars and the minor planets Flora, Juno, Iris, Victoria and Sappho were used and

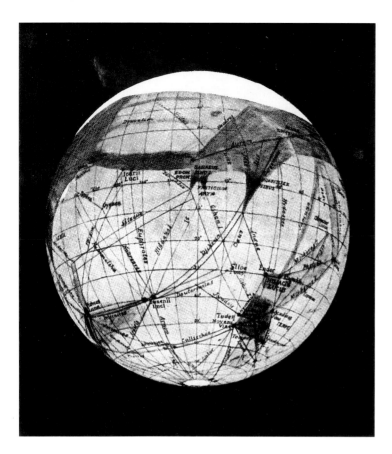

Above: **The planet Mars as mapped in 1905 with white polar caps, the dark, so-called continental areas crossed by** *canali* **(canals), and the two small satellites, Phobos and Deimos.** *Below:* **The Equatorial Telescope at Washington, DC.**

the derived distances ranging from 92 to 93 million miles. The necessary observations were made in the year 1872 (Flora), 1874 (Juno), 1877 (Mars), 1888 (Iris), and 1889 (Victoria and Sappho). An active worker in many of these investigations was Sir David Gill, who held the title of His Majesty's Astronomer at the Cape of Good Hope.

Other methods of a less direct description were also used, such as by carefully measuring the displacement of stars by the aberration of light, combined with the actual velocity of light found by physical laboratory experiment. The displacement of a star by aberration depends upon the ratio of the velocity of light to that of the Earth in its orbit around the Sun. The latter velocity can therefore be derived from the two measurements referred to, and the length around the elliptical path, and hence mean radius of the Earth's orbit, the quantity required, can be calculated.

During the half century or more after Schroeter, little was done in observation in terms of planetary Astronomy, but from his results in 1800, Bessel inferred a rotation period for the planet Mercury not much different from the Earth's. Little reliance was placed on this value, although many years later, in 1882, **WF Denning (1848-1931)** found that it accorded with several days' observations. In that year, systematic study of Mercury was undertaken by Schiaparelli at Milan, the scrutiny of the planet being in daylight, as evening or morning observations had to be at low altitudes in the sky, owing to the small angular distances from the Sun, with consequent great atmospheric disturbances. Schiaparelli kept continuously observing, hour by hour and found faint markings that did not change their positions on the disc, indicating a much longer period than Schroeter's. In 1889, he announced that Mercury rotated in a period of 88 days, which is equal to its period of orbital revolution, and that it therefore kept the same hemisphere of its surface turned to the Sun. These conclusions were supported by **Percival Lowell (1855-1916)** at Flagstaff, Arizona, in 1896. To both observers, Mercury appeared to have little or no atmosphere.

In 1788 and 1811, Schroeter had found a short rotation period of 231 hours for Venus, and in 1841 **F de Vico (1805-1848)** thought he had confirmed this, although William Herschel had never been able to see any of the markings on which Schroeter had based his result. Schiaparelli studied Venus from 1877 to 1890, in the daytime as with Mercury, and found it had a rotation period of 225 days, equal to its period of revolution around the Sun. Several other observers supported this theory, including Lowell. An attempt to settle the matter with the spectroscope made in 1900 by Belopolsky at Pulkowa, gave displacements of the spectral lines which agreed with the short period. However, the results were experimental, with apparatus not yet competent to deal with such a question, as even with the rapid period, the line displacements would be only those corresponding to a velocity of about half a mile per second.

Many other observers of Venus thought that they found evidence of the more rapid rotation period. But the markings on the disc of the planet, which is very bright naturally, are of a vague and changeable kind, suggesting a cloud-laden atmosphere. The presence of an atmosphere was also indicated by the observed prolongation of each of the points of its image when of a thin crescent shape. This had been noted by several, including Madler with the 9.6-inch refractor at Dorpat in 1849, who had seen them extended for 30 degrees beyond the semicircle. CS Lyman in 1866 and HN Russell in 1898 had observed these carried right around the complete disc, Venus

Above: **Asaph Hall observed two moons orbiting Mars in 1877.**

being then only a degree or two from the center of the Sun. The first to note similar evidence of an atmosphere seems to have been D Rittenhouse of Philadelphia, who observed at the entry of the planet on to the Sun during the transit of 1769, that part of the edge of the planet off the solar disc appeared illuminated so that the whole outline of the planet was visible.

The reported visibility to several observers of the dark part of Venus—similar in appearance to the Earthshine on the Moon—during daytime or in bright twilight, and seldom if ever at night, appears to have been the result of some optical illusion, as such an appearance should be much more easily seen after dark when it would be less liable to effacement by the light of the sky.

The essential permanence of the dark markings seen on the disc of Mars was established by Beer and Madler by their observations from 1830 to 1839. Before that time, Schroeter and others (but not Herschel) had considered them to be changeable and cloud-like. In 1862, J Norman Lockyer confirmed Beer and Madler's conclusions. From observations that year, several drawings of William Herschel, and one by Huygens in 1672, **F Kaiser (1808-1872)** derived a rotation period of 24 hours, 37 minutes, 23 seconds. This was about two minutes shorter than that found by Herschel and identical to what we now know to be true. The seasonal fluctuation of Mars' white polar caps, discovered by Herschel, was confirmed. The opinion gained ground that Mars is most similar of all the planets to the Earth in physical constitution.

At the favorable opposition in 1877, Schiaparelli, using an 81-inch refractor at Milan, began an intensive study of Mars, which lasted for the next 13 years. In the first year of his work, he found that the brighter so-called continental areas, were crossed by dark linear markings, to which he gave the Italian name of **canali** or **channels**, which unfortunately translated into English as **canals**. The most prominent of these had evidently also been seen by Beer and Madler, Dawes and others, but had been viewed merely as straits. In 1879, Schiaparelli found that one canal appeared as a double marking, and in 1882 he noted the doubling of several more. These discoveries met with considerable skepticism, but other observers at Nice, using a 30-inch refractor, and at Lick with the 36-inch telescope, confirmed Schiaperelli's findings. Even users of smaller telescopes saw many of the canals. In 1892, William Pickering, and in 1894 AE Douglass, found that they were not confined to the areas termed 'continents,' as some crossed the darker parts which had been thought to be seas, a phenomenon also noted in 1888 by Perrotin of Nice in regard to several of the continental canals.

In 1894, Percival Lowell founded his observatory at Flagstaff, Arizona, with the principal equipment at first of an 18-inch refractor, and later a 24-inch refractor and a 40-inch reflector, to study planets, Mars in particular. Lowell and his colleagues believed their observations showed narrow canals in great number. Because all of the 'seas' were crossed by canals, they therefore could not really be bodies of water. In fact, these darker areas, according to the Lowell observers, underwent seasonal changes that suggested vegetation.

It seemed reasonable that the canal markings had an origin, although some users of very large telescopes believed what was seen, at or near the limit of vision, was something really much more intricate in detail. Lowell's fascinating writings give a modern day reader the impression he was trying to make a case for the existence of a combination of features which was actually a huge irrigation system that was the product of intelligence, although there is a suspicion of some bias in such observations. It should be noted, however, that the same scale photographs, taken at Flagstaff, Lick and elsewhere, certainly showed one or two larger canals, exactly where they are observed visually, as markings of a linear nature.

Evidence of cloud formations, probably mist or dust, was noted in 1873, 1888, 1890 and 1896 by different observers, but the general view was that the Martian atmosphere was thinner than that of the Earth. Spectroscopic observations by Huggins in 1867, and **HC Vogel (1842-1907)** in 1873, seemed to show the presence of water vapor, although **WW Campbell (1862-1938)** in 1894, and Keeler in 1896 at Lick Observatory, did not confirm this.

While observing Mars from the US Naval Observatory in Washington, DC, the American astronomer **Asaph Hall (1829-1907)** had determined on 11 August 1877 that the red planet was accompanied by two tiny moons. Named Deimos (terror) and Phobos (fear) after the characters of (some sources say 'the horses of') the Roman war god Mars, the two moons are irregularly shaped rocks pocked with numerous craters. Phobos, the larger of the pair, is just over 17 miles in length, while Deimos is less than nine miles long. Because of their shape, size and texture, it is thought that they originated among the asteroids and became trapped in Martian orbit at the time of the formation of the Solar System. Neither Deimos nor Phobos have the mass to allow them to hold an atmosphere, but they exert sufficient gravity to retain a thin layer of dust on their surfaces—which is perhaps residue from the meteorite impacts that caused the cratering.

The discovery of asteroids, 10 of which were known before 1850, proceeded at an accelerating rate in the later part of the nineteenth century. By 1870, 110 had been detected and all but one had been given a satisfactory orbit, a number, and (by the discoverer) a name. By the beginning of 1900, 559 asteroids had been found, and 452 given orbits, numbers and names. A century later, the number was well past 3000 and still growing.

Photography, as a means of detection, was introduced in 1891 by **Max Wolf (1863-1932)** of Heidelberg. At the end of the century, he had found more than 100 asteroids, Charlois of Nice another 100, Palisa of Vienna more than 80 and CHF Peters over 50.

Among these discoveries, the object of chief interest and astronomical value was the asteroid detected by Witt of Berlin in 1898, which he named Eros. Its mean distance from the Sun was the smallest known for an asteroid—136 million miles. However, it had a very eccentric orbit, and it passed within 14 million miles of the Earth when it had a very large parallax (60 seconds). Unfortunately, one of these favorable oppositions occurred in 1894, several years before its discovery, but the value of this asteroid for determining the scale of the Solar System in the future, and hence the solar parallax, was at once obvious.

In 1866, when the number of asteroids known was only about 90, and 10 years later when nearly twice as many were available for study, D Kirkwood remarked that their distribution in distance from the Sun contained marked gaps. He suggested that these comparatively unoccupied spaces are just where the periods of revolution around the Sun are connected by a simple ratio with that of Jupiter. Such gaps are found where the period is one-third, two-fifths, one-half, and three-fifths of Jupiter's, a place where the perturbations by Jupiter would be cumulative, keeping the spaces clearer than other regions. Kirkwood also suggested a similar explanation for the Cassini division in Saturn's ring system, one of the satellites, Mimas, being chiefly responsible.

Study of Jupiter and its cloud belts during the nineteenth century led to the hypothesis that it and the other giant planets are in a state between the solar and terrestrial stages of development, possessing hot interiors and very warm surfaces. This hypothesis was based on the movements and changes in Jupiter's belts and the fact that its surface markings had a quicker period of rotation near the equator, as is found on the Sun. JD Cassini and Schroeter both had observed this peculiarity. Their measurements of the amount of light received from Jupiter seemed to indicate that there was probably a certain amount of emitted light, the results of Zollner in 1865 and Muller in 1983 giving **albedos**—an albedo of a spherical body being the ratio of the total amount of sunlight reflected from the body in all directions, to that which falls on the body—nearly equal to that of white paper or of fresh snow. Observers such as Barnard, Keeler, Lowell and Denning, reported changes and relative movements, as well as colors, in the belts, which all appeared to be consistent with the 'semi-sun' hypothesis, strongly advocated by **RA Proctor (1837-1888)** in his writings.

Another striking feature on the planet's disc, which apparently conformed to this theory, was the **Great Red Spot**, first noted by Schwabe in 1831, although it has been suggested that Robert Hooke and JD Cassini may have seen it in the seventeenth century. A marking on the planet's surface seen in 1676, which Roemer thought might be used in place of the occultations of Jupiter's moons for measurement of the velocity of light, may have been the Great Red Spot. The changeable phenomena of the transits of satellites and their shadows across the disc of Jupiter were also thought by some

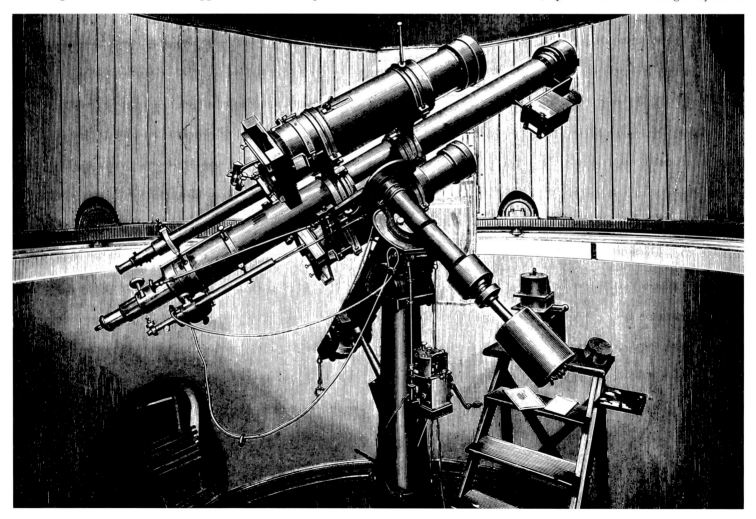

THE SCIENCE OF ASTROPHYSICS

Left: **Max Wolf designed this photographic equatorial telescope which revolutionized minor-planet search techniques.**
Above: **The Lick 36-inch telescope revealed Jupiter's fifth moon.**

to lend support to the hypothesis of great heat. However, the general view of the structure and origin of the Great Red Spot was that it was, in Lowell's words, 'a vast uprush of heated vapor from the interior.' Others attributed a vortex form to it. Later research showed the great heat hypothesis to be false.

The spectroscopic observations of Huggins between 1862 and 1864 and those of Vogel between 1871 and 1873, showed a solar spectrum with bands in the orange and red, but no explanation of these bands was possible by the science of the time. Photography of Jupiter was undertaken between 1890 and 1892 at Lick Observatory and several years later at Paris, but the results, although initially promising, had little more than pioneer value. These conspicuously show the Great Red Spot but the main belt arrangement is without much detail.

In 1892, Edward Barnard added a fifth satellite—little Amalthea—to the Galilean four. He discovered Amalthea inside their orbits. It had a period of nearly 12 hours, or only about two hours greater than the rotation period of Jupiter itself, revolving at only one to five times the radius of the planet away from its equatorial surface.

Asaph Hall discovered a bright spot in the equatorial regions of Saturn in 1876. From it, he calculated it had a rotation period of slightly under 10.25 hours, or about two minutes shorter than had been noted by Herschel from a similar marking in 1794. Between 1891 and 1893, several such spots were observed, yielding the same or slightly shorter periods, a difference no doubt due to individual motion of the spots similar to what had been seen on Jupiter. From his own observations of a number of spots, Stanley Williams found a continuously decreasing rotation period of the equatorial regions throughout the years 1891 to 1894.

In 1898, William Pickering added to the eight moons of Saturn found by Huygens, Cassini, Herschel, and Bond, by the discovery, photographically, of a faint satellite revolving well outside the orbits of the others. Named Phoebe, this satellite is peculiar in that its motion around Saturn is in a direction opposite to that of the others. Phoebe continues to be the outermost known moon of Saturn.

Even today, Uranus is a difficult object for ground-based observation. In 1883, with the 23-inch refractor at Princeton, Young noted dusky bands, and in 1884, observers at Paris, using a 23.5-inch Coudé refractor, also saw two bands. Similar results followed in the same year with the 30-inch refractor at Nice. Observers there also noted a bright spot near the equator, indicating the planet had a rotation period of about 10 hours, which is not much different from the period found spectrographically nearly 30 years later. Uranus' elliptic shape was evident to many users of large instruments.

The spectrum, as seen by Huggins, Vogel and Keeler, showed broad bands more conspicuous than either Jupiter or Saturn. Two new satellites were identified by William Lassell in 1851 with his two-foot aperture reflector, bringing the total known for Uranus to four, as only two of the six which William Herschel thought he had discovered had been verified. The planes of the orbit of the four were noted at 82 degrees to the plane of the ecliptic, the movement being in a direction opposite to that usual for planets and satellites. Indeed, little more would be known about the seventh planet until the Voyager spacecraft arrived in 1986.

Neptune is, of course, an even more difficult telescopic object, its observable disc being little more than half the size of Uranus, and it is even less brightly illuminated by the Sun's light. The Sun's light on Uranus is 1/360th as intense as on the Earth; on Neptune it is only 1/900th.

In 1899, TJJ See believed he saw equatorial belts, using the 26-inch Washington refractor, but no definite markings were visible to Barnard with either the Lick 36-inch or the Yerkes 40-inch refractors.

As far as could be judged from the faint spectrum visible to Huggins and others, its character was of the same type as that of Uranus. No satellites additional to the single one found by Lassell in 1846 were added until Gerrard Kuiper found Nereid in 1949. The American Voyager spacecraft would add five more to the list in 1989.

During the second half of the nineteenth century, more than 200 comets were found, eight of them classed as 'bright.' Some of the most spectacular of these were Donati's of 1858 (specially interesting from the beauty of its curved tails and the brightness of its head); Tebbutt's of 1861 (the Earth passed through its tail on 30 June 1861, when some observers believed they noticed a diffused glare in the sky); Coggia's Comet of 1874 (with a tail extending more than 45 degrees in the sky); the Great Comet of 1882 (visible in daylight, and seen telescopically to throw off a satellite comet); and the comet of 1887 (first seen by a farmer near Cape Town, it had no head or condensation). All of these comets had striking tails millions of miles long and orbits of a greatly elongated ellipse form, with periods of hundreds of thousands of years, and parabolic in shape. From a comparison of recorded descriptions, the three most brilliant comets of the nineteenth century were those of 1811, 1843 and 1858. The 1843 comet was the brightest, and that of 1858 the most gracefully shaped.

Left: **The first American heliometer, erected in 1882.**

PROGRESS IN ACCURACY OF MEASUREMENT

Of the accuracy of measurements, on which so much in Astronomy depends, it is of interest to note the progress from the earliest times. The average errors may be taken to have been as follows: Hipparchus (second century BC), four minutes of arc; Tycho Brahe (sixteenth century), one minute; Hevelius (seventeenth century), one minute or less; Flamsteed (late seventeenth century), 10 seconds of arc; Bradley (eighteenth century), two seconds; Bessel (heliometer, early nineteenth century), two-tenths of a second; first photographs (mid-nineteenth century), one-tenth of a second; twentieth century long-focus telescope photographs, one-fortieth of a second. The improvement in accuracy shown is nearly 10,000 times.

An example of the extraordinary accuracy of modern observations is provided by the discovery and measurement of the **variation of latitude**. This was found from change in the astronomically determined value of latitudes of certain observatories, the amounts being much less than half a second. Klistner of Berlin in 1888 and SC Chandler of Boston in 1891 were the first in this field. The phenomenon is explained by a small change in the position of the Earth's axis of rotation with respect to the Earth itself, probably due to seasonal changes in ice and snow, vegetable growth and barometric pressures. The movements of the pole on the Earth's surface are contained in a circle of less than 50 feet radius.

In 1871, Zollner revised Olbers' theory about the electrical formation of comets' tails, and in 1874 **T Bredechin (1831-1904)** of Pulkowa, classified the tails of comets into three fairly distinct types according to the strength of the repulsive force from the Sun, with the degree of curvature being inversely proportional to this force. These three types were clearly recognized for a large number of comets, and further experience confirmed the substantial accuracy of his classification, with the suggestion of even more powerful repulsive forces in some cases.

The first spectroscopic observation of a comet was made in 1864 by **G Donati (1826-1873)**, who discovered the great 1858 comet. Donati found three bright bands of yellow, green and blue, and in 1866, Huggins and Secchi noted that Tempel's Comet had three bands, as well as a continuous spectrum. In later comets, dark lines of the solar spectrum were seen, thus indicating reflected solar light as well as bright radiation from a glowing gas or gases. In 1868, Huggins identified these same three bands in Winnecke's Comet as being due to a hydrocarbon.

In a few comets, the hydrocarbon bands were not present, although the majority of comets had them. For instance, a comet discovered in 1892 by E Holmes, a London amateur, had a faint continuous spectrum only, as did Brorsen's Comet in 1868 and another comet noted in 1877. In 1882, the D line of sodium was seen in the spectrum of Wells' Comet, and later it was also visible in the spectrum of the Great Comet of that year, in both cases when the comet was near the Sun. This led to the discovery that, for comets which pass close to the Sun, the order of spectrum is first a hydrocarbon one. As the comet continues to approach the Sun, sodium lines appear. Finally, a number of other bright metallic lines, including iron, appear, a continuous reflected solar spectrum always present as well.

The practice of 'comet-hunting' or 'comet-seeking' of which Pons, Messier and others were notable exponents, died away after the first quarter of the nineteenth century, but was revived again about 1880, when Brooks, Barnard, Perrine and Swift became successful searchers.

The first comet to be deliberately photographed was in 1881. Discovered by Tebbutt, the photographs were taken by Janssen and Draper. Another one was accidentally photographed in the vicinity of the totally eclipsed Sun in 1882. In the same year, the Great Comet was photographed at the Cape Observatory on a background crowded with stars, thus showing the full possibilities of stellar photography.

In 1891, **FF Tisserand (1845-1896)**, Callendreau, and H Newton independently published the suggestion that certain comets had been captured in the Solar System, the gravitational pull of the larger planets Jupiter, Neptune, Uranus, and Saturn having changed their parabolic orbits into elliptical ones. Some time before this, RA Proctor advocated the hypothesis that certain particular comets had been ejected from the large planets.

In 1865, Hoek of Utrecht drew attention to a different kind of comet grouping. The most noteworthy group was made up of the long-period comets of 1668, 1843, 1880, 1882 and 1887, all of which moved in space in greatly elongated elliptical orbits, of similar shape and disposition, that brought them very near to the Sun at perihelion. Members of such a group were regarded as fragments of a larger, earlier comet which, like Biela's Comet, had broken into parts which then separated out along the original orbital path. A marked tendency to disintegration into parts observed in the comet of 1882

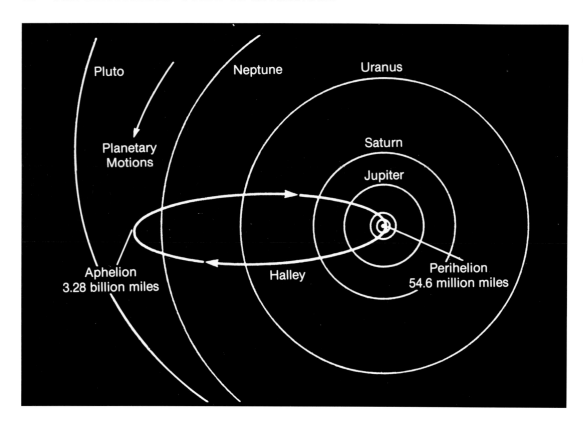

Left: **The highly elliptical orbit of Halley's Comet brings it very close to the Sun at perihelion.**

seemed to lend some support to the hypothesis for the group of which it was believed to be a member.

The connection of certain meteor showers with periodic comets has been mentioned earlier. In the case of the Leonids, the next great shower was expected for 1899, but Johnstone Stoney and **AM Downing (1850-1919)** projected that perturbations of the main swarm of meteors by Jupiter and Saturn would so interfere with their orbit as probably to cause the richest part to miss the Earth, passing about two million miles from it. As a consequence, there was no great display in 1899 or 1900, much to disappointment of the public. A somewhat better display, visible in America, occurred in 1901, probably from a part of the swarm well away from the rich main body which had provided the great showers of 1799, 1833 and 1866.

The existence of swarms of meteors, although not identified with particular comets, was made evident by discovery of the great number of radiant points, or points of the sky from which meteors appeared to radiate on a particular night—or series of several nights—a year. In 1867, **E Heis (1806-1877)** of Munich published a catalogue giving the positions in the sky and dates of activity of 84 such points, and Schiaparelli issued another catalogue in 1871, listing 189 points. However, the greatest observer of meteors of the time was the English amateur, WF Denning, whose work extended into the first third of the next century. He claimed to have discovered **stationary radiants**, or radiants that did not appear to alter their position in the sky from night to night, as should be the case owing to the change of direction of the Earth's motion in its orbit, except with meteors of very high velocity.

The reality of stationary radiants has been the subject of much controversy and various attempts at explanation. Recent expert opinion appears to be that they have no real existence as a general feature of meteor radiation, such as Denning claimed, although one may appear to exist in a restricted region of the sky for special reasons.

A well-marked shift of the radiant point in accordance with the Earth's changing direction of movement was observed for the August Perseid Shower. Le Verrier had predicted this should be the case. It was later noted that the radiant was not the same point source from night to night but appeared to have an elongated shape extending from Perseus into Cassiopeia. In 1877, Denning made a series of observations between 19 July and 10 August and found that the radiant moved eastward at the rate acquired by the Earth's motions. Although this observed night to night shift seemed at first to be too great, mathematical analysis—making allowances for the Earth's attraction on the meteors—showed that the observed radiants for different nights belonged to a compact group. Denning later confirmed this shift of the Perseid radiant with greater precision, and, in addition, he later detected a similar movement in the radiants of two other showers—the Lyrids and the Geminids. Two results of promise for the future were: the photographing in 1897 of the spectrum of a meteor at the Harvard Observatory Station in Arequipa, and the determination by Elkin in 1898 of the path and the orbit of a Leonid meteor from photographs at Yale College.

In addition to the *Fundamenta Astronomiae* based on Bradley's work, Bessel brought up the number of catalogued stars to more than 50,000 by a series of over 75,000 observations. His assistant and successor, WH Argelander published between 1857 and 1863 the *Bonn Durchmusterung*, a catalogue, with an accompanying atlas, of 324,198 stars in the northern sky. This work was extended between 1875 and 1885 to cover part of the southern hemisphere of the sky by a catalogue of 133,659 additional stars by **E Schonfeld (1828-1891)**. Photographic observation of the rest of the southern sky, made under the direction of D Gill at the Cape of Good Hope, completed the entire heavens nearly to the tenth magnitude, with positions reliable to about a second of arc. The work of preparing the catalogues, based on the observations being carried out, was finished in 1900 by **JC Kepteyn (1851-1922)**.

Arising from this work, an international congress held in Paris in 1887 decided to undertake a photographic survey of the entire sky by chart and catalogue. The work was to be shared among 18 observatories, from Helsingfors in the northern hemisphere to Melbourne, Australia. The chart was to include stars down to the fourteenth magnitude and the catalogue down to eleventh magnitude. Each observatory was to use the same instruments—13-inch aperture photographic refractors with 11-foot focal length. Each plate covered four square degrees of the sky—about 18 times the Moon's area—and as there had to be duplication to check any possible error, 22,000 of them were required.

Invented by Hale in 1924, the **spectrohelioscope** provided a view of the solar surface, through an eyepiece, in monochromatic light, as the spectroheliograph provided a photograph. The two slits, which moved across the Sun's image, in the spectroheliograph were made to oscillate to and fro. This yielded a series of visual impressions of neighboring narrow strips of a part of the Sun's image which, (as in television), become a single continuous picture owing to persistence of vision. The monochromatic image is usually that of the red hydrogen line, as other lines, useful in the spectroheliograph, are in the short-wave blue or violet region of the spectrum to which the eye is less sensitive. The instrument is ideally suited for observation of the disturbed regions of the Sun's surface over and around spots, and it can also be adapted to show the line-of-sight velocity of any bright hydrogen cloud through the observed displacement (by Doppler's principle) of the image to the red or blue side of the spectrum which accompanies such a movement.

Several forms of reflecting telescopes for photography evolved. The first and best known was described in 1930 by a German amateur, B Schmidt. The Schmidt reflector consisted of a spherical mirror (a more easily made form than the paraboloid of the ordinary reflector) with a thin (and rather difficult to make) correcting glass plate of varying thickness set in front of the mirror. This plate corrected the paths of the light rays so that good images were given over a very much larger field of view than had been possible with the orthodox form of the reflector, some of the correction being provided by the photographic film's attachment to a holder with a specially curved surface. This and several related forms of instrument were exceedingly powerful in discovery of faint objects or faint extensive nebulae, and the definition was so good that the original plates may be magnified considerably to study detail. By the mid-twentieth century, there were more than 30 Schmidt cameras in existence, ranging from 10 to 60 inches in aperture.

Improvements in instruments for stellar photometry, such as the selenium cell and the photoelectric cell, effected remarkable increase in accuracy. The selenium cell was first used by Stebbins in 1910, and the photoelectric cell by Guthnick in 1913, on variable stars. They produced results correct to a hundredth of a magnitude or less, against the tenth or so possible to eye estimation and the twentieth to ordinary photometers. By means of these special photometers, minute fluctuations in the light of spectroscopic binaries were detected. One of the first was by Stebbins when he showed that Algol had a secondary minimum between its principal ones, proving that the fainter component had an appreciable brightness of its own which was lost when it moved behind its primary.

In 1931, B Lyot of the Meudon Observatory in France ascended to the summit of the Pic du Midi, 8500 feet above sea level, to photograph the Sun's corona without an eclipse. Huggins had tried to do this in London as far back as 1882 but without success. Lyot's instrument has been given the name **coronagraph**. Its chief features were a disc blotting out the Sun itself, an object glass free from scratches and dust and a monochromatic light filter.

The dust-free atmosphere at the altitude where Lyot took the photographs was also a necessary condition. Two other stations for coronagraph work were soon built, including Harvard's at Climax, Colorado at 11,500 feet, and one in the Alps, established by Zurich University. With the improvement in the new instrument, more information for the inner corona and the Prominences were available in a few years, much more quickly than could have been collected at scores of eclipses.

With all of the equipment in place and coming on line at the dawn of the twentieth century, an exciting new age of discovery was beginning. The first, and one of the most important of the twentieth century's major ground-based astronomical discoveries was that of a ninth planet.

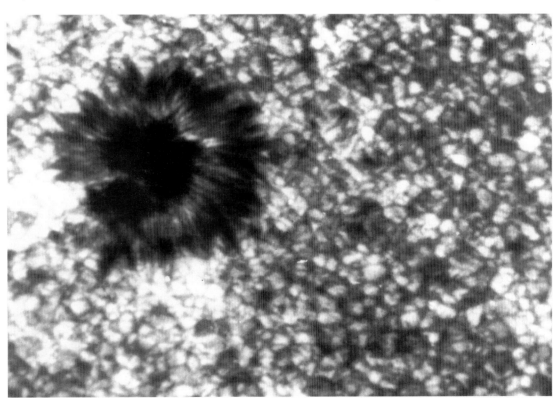

Right: **This photograph of a small sunspot, roughly the size of the contiguous US, was taken from the Space Shuttle *Challenger* in 1985, using a spectrohelioscope.**

THE DISCOVERY OF PLUTO

During the mid- to late nineteenth century, astronomers studying the revolutions of Uranus and Neptune detected slight anomalies that could be explained only by the gravitational effect of another body farther out in the Solar System. Around the turn of the century Percival Lowell took up a systematic search of the heavens, looking for what he called 'Planet X.' When Lowell died in 1916, others continued the search, including William Pickering of Harvard, who called the yet-undiscovered object 'Planet O.'

In 1915, and again in 1919, Pluto was actually *photographed but not noticed* because it was much fainter than it had been predicted to be. By this time, the organized search for Planet X was largely abandoned. In the meantime, Pickering altered his theory regarding the hypothetical location of Planet 'O,' and for the first time predicted that the perihelion of its orbit might actually bring it briefly closer to the Sun than Neptune. It was a radical idea that turned out to be accurate for Pluto.

In 1929, the Lowell Observatory at Flagstaff, Arizona, resumed the search begun by its founder, using a 13-inch telescope and a wide-field survey camera. This proved to be the right approach, and on 18 February 1930 the young astronomer **Clyde Tombaugh (1906-)** identified a new planet in some photographs he had taken the previous month. The discovery was announced a month later on the 149th anniversary of the discovery of Uranus, and the new planet was called Pluto after the Roman god of the dead and the ruler of the underworld. The name was considered appropriate because of the planet's enormous distance from the Sun's warmth, and also because the first two letters were Percival Lowell's initials.

In the first years after it was discovered, physical data about Pluto was virtually impossible to obtain. In 1950, however, Gerrard Kuiper at the Mount Palomar Observatory estimated its diameter at 3658 miles, making it the second smallest planet in the Solar System. In 1965, it was observed in occultation with a 15th magnitude star, confirming that its diameter could not exceed 4200 miles. Thus it was that the 3658 estimate held until the 1970s. In 1976, methane ice was discovered to exist on Pluto's surface. Until then the planet's faintness had been attributed to its being composed of dark rock. Since ice would tend to reflect light more than dark rock, it would follow that if it *were* 3658 miles in diameter *and* covered with methane ice, it would be brighter than it is. Therefore, it was decided that Pluto was smaller than originally suspected, leading us to conclude that its diameter is less than the 2160 mile diameter of the Earth's Moon, and probably as small as 1375 miles. This would make it the smallest of the nine planets and smaller than *seven* of the planetary moons.

It has also been suggested that Pluto is perhaps the largest of a theorized belt of trans-Neptunian asteroids. However, that notion fails to take into account that Pluto is two and a half times the diameter of Ceres, the largest known asteroid, and nearly seven times larger than the average of the 18 largest known asteroids. Among the arguments that *can* be made for its not being a planet, or at least for its not being a 'normal' planet, are the peculiar aspects of its behavior. As we have noted, it has an extremely elliptical orbit. This orbit ranges from an aphelion of 4.6 billion miles to a perihelion of 2.7 billion miles. The latter is actually closer to the Sun than the perihelion of Neptune's much more circular orbit, as Pickering

Below: A diagram of Pluto's orbit. *Opposite:* These photographic plates taken on 23 and 29 January 1930 of the constellation Gemini show the trans-Neptunian planet Pluto.

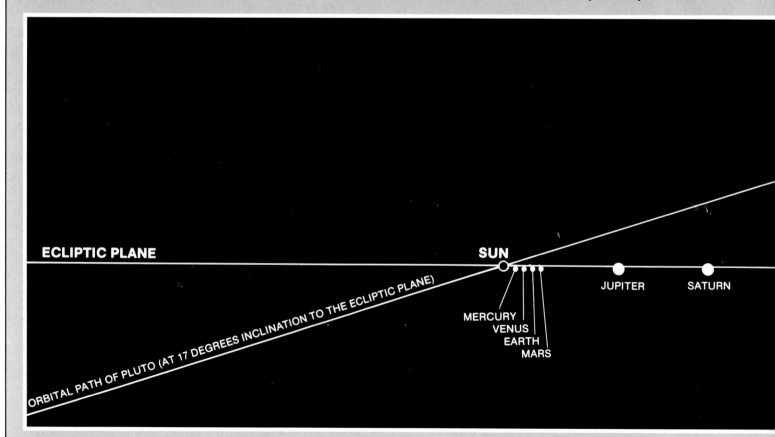

had predicted. It has been pointed out that this highly elliptical orbit is more characteristic of asteroids.

A second aspect of Pluto's behavior that sets it apart from other planets is its steep inclination to the elliptic plane. The orbits of all the planets are within 2.5 degrees of this same plane, except Mercury, which is inclined at seven degrees, and Pluto itself, which is inclined at an acute 17 degrees, making it very unusual among its peers.

A theory concerning the physical nature of Pluto holds that at one time it was actually one of the moons of Neptune. It is further theorized that Pluto was somehow thrown out of its Neptunian orbit by some calamitous interaction with Neptune's moon Triton—perhaps even a collision. One of the Solar System's largest moons, Triton is more than twice the size of Pluto and, as such, might have had the gravitational force to slam a competing object out of Neptunian orbit if it ventured close enough. Both Triton and Nereid have unusual orbits that might possibly be relics of such a colossal event.

While its behavior partially defines it, and certainly sets it apart from other planets, less is known about Pluto's physical characteristics than is known about any other planet. Since no spacecraft will visit it in the twentieth century, we are left with only educated guesses about Pluto. We know that it is extremely cold, with noontime summer temperatures rarely creeping above -350 degrees Fahrenheit. Its rocky surface is known to contain methane, probably in the form of ice or frost. Water ice may also be present, though this is not likely, and Pluto's mass suggests a rocky core. Pluto has generally been thought to have no atmosphere because its relatively small mass wouldn't give it sufficient gravity to retain an atmosphere, and it is too cold for even such a substance as methane to easily exist in its gaseous state. However, Scott Sawyer of the University of Texas has discovered what may be a tenuous methane vapor atmosphere on Pluto.

The discovery of the Plutonian moon Charon came about indirectly in 1978. While **James Christy** at the **US Naval Observatory** in Flagstaff, Arizona was attempting to measure Pluto's size, he thought he'd noticed that it was not spherical. Further observations led him to the conclusion that the elongation he had observed was due to the presence of a satellite very close to Pluto. Further calculations indicated that this newly discovered body was as close as 10,563 miles from Pluto.

THE GREAT OBSERVATORIES

Prior to the nineteenth century, most major observatories, such as the Royal at Greenwich and the US Naval in Washington, were located near sea level. A development of primary importance in the nineteenth century was the practice of selecting only high-level sites for new observatories, as was the case with the **Lick Observatory** on Mount Hamilton, California, at 4200 feet above sea level, and the **Lowell Observatory** at Flagstaff, Arizona, at an elevation of 7310 feet. There was also an increasing desire to choose places where the amount of clear sky was great and steady air conditions prevailed. This practice has been continued—although not quite as often as could be desired—and careful choices have been made even more desirable by the growing need for observation points not affected by the sky illumination of modern towns, and the consequent limitation in photography of faint objects because plates fog up.

At the end of the nineteenth century, the largest refractor was the 40-inch at the **Yerkes Observatory** at Williams Bay, Wisconsin, built by the University of Chicago in 1897. In 1900, there were about 20 refraction telescopes of 24 inches or more in aperture, four of them having object glasses corrected for photography. Until 1935 the biggest reflectors were a 72-inch at Victoria, British Columbia (1919) and the famous 100-inch at Mount Wilson, California (1917). The maximum practicable for the refractor appears to have been reached with the Yerkes giant. This was due to several factors, such as the problem of support against flexure (an object glass cannot be supported at its back as reflectors can), the difficulty of obtaining very large, optically perfect transparent glass discs, and the greater cost compared to that of a reflector of equal, or even greater, light-grasping power. The achromatism of a reflector, which brings rays of light of all colors to the same focus, is a great advantage. One improvement, introduced in the 1930s, substituted aluminum for silver as the reflecting coating on the surface of mirrors. The advantages were a longer time before re-coating (years against months); superior reflectivity for the violet and ultraviolet regions of the spectrum, with shorter exposures necessary; and greater powers of photographing the spectrum in the ultraviolet beyond the range of a silver-on-glass mirror.

However, early twentieth century telescopes of new types were being developed during the century. One of them was the **Tower Telescope** for solar work, the largest of these being the 150-foot one at Mount Wilson. The rays of the Sun arriving at the top of the 150-foot tower were directed to a 12-inch object glass by a reflecting system of two flat mirrors 24 inches in diameter, one of which was driven by a mechanism to follow the Sun and keep its rays centered on the second mirror. The light then passed through the object glass and went down the shaft of the tower to form a 17-inch diameter image of the Sun at the ground level, thus providing a large-scale picture of the visible activities on the solar surface, including the sunspots. A drawing is made every clear day showing the size and form of each spot and its position on the Sun's disc, as well as other observations, useful in the study of the Sun's activity.

In the field of **Optical Astronomy**, the era of the big refractors was over at the beginning of the twentieth century. The new century would be the era of the reflector telescope. Three moderate-sized refractors (26 to 27 inches) were built in South Africa between 1925 and 1928, but nothing would top Yerkes, and there would be no more big refractor installations.

The tide had turned to reflector telescopes and the desire to start building observatories on remote mountaintops where the air was clear, the atmosphere thinner, and there was less

Below: The Lick Observatory. Right: The Palomar Observatory.

interference from the lights of surrounding cities. The first of the great reflectors was the Hooker Telescope on Mount Wilson in Southern California, the brainchild of **George Ellery Hale (1868-1938)**, the man who had convinced Chicago millionaire Charles Tyson Yerkes to underwrite the huge University of Chicago refractor. This time, Hale turned to California businessman John D Hooker, who opened his pocketbook for the construction of the observatory on Mount Wilson, where the 100-inch **Hooker Telescope** that was installed remained state-of-the-art for over three decades.

For the 30 years after the Hooker Telescope began scanning the heavens from Mount Wilson in 1917, there was little in the way of major observatory construction in the world. The **McDonald Observatory** opened in Texas in 1939 with an 82-inch telescope, but there would be nothing to top the 100-inch monster on Mount Wilson until after World War II.

George Ellery Hale was not idle during this time, however. Spending as much time lobbying philanthropists as he did scanning the firmament, Hale next turned to the Rockefellers for funding. This time, what he had in mind was nothing short of a reflector telescope of 200 inches, double the size of the Hooker giant. Hale's ultimate dream was eventually fulfilled, but, sadly, not until a decade after his death when, in 1948, a 200-inch reflector finally opened at the **Palomar Observatory** on 5600-foot Mount Palomar in California. Hale's posthumous achievement was officially designated as the **George Ellery Hale Telescope**.

The Hale telescope remained the largest in the world for another three decades until the Soviet Union built the 234-

ASTRONOMICAL SOCIETIES

Throughout the nineteenth century and into the next, there was a proliferation in the formation of astronomical societies. These included the Royal Astronomical Society (1820), the Astronomische Gesellschaft (1863), the Societe Astronomique de France (1887), the Astronomical Society of the Pacific (1889) and the British Astronomical Association (1890). In Canada, the Royal Astronomical Society of Canada (1890), and (for professionals) the American Astronomical Society in the United States, were important societies.

Soon there were organizations of varying size and importance in Belgium, Czechoslovakia, Denmark, Greece, Holland, Italy, New Zealand, South Africa, Soviet Russia, Sweden and Tasmania. A fully comprehensive list would be very long indeed. In the United States, for example, there was a society especially for the study of variable stars and also one for meteors, and there are said to be 40 or more local societies which have reached some degree of stability and permanence, including a number for mutual help in telescope making.

Opposite: **The 200-inch Hale Telescope.** *Below:* **The Siding Spring Observatory's 40-inch telescope** *(foreground)* **and the 154-inch Anglo-Australian Telescope** *(background)*.

Above: The telescope at Cerro Tololo Inter-American Observatory, Chile is the largest optical telescope south of the equator.
Below: The Multiple Mirror Telescope at Whipple Observatory.
Opposite: A diagram of the Nicholas Mayall Telescope.

inch **Bolshoi Teleskop** on 6900-foot Mount Pastukhov near Zelenchukskaya in what is now the Russian Republic. Though the Russian telescope was larger than the Hale in size, it was considered much inferior in optical quality, and the Hale remained the best in the world.

The decade between 1969 and 1979 is remembered as an important one in the history of Optical Astronomy for it saw the construction of no fewer than nine reflector telescopes with larger apertures than the one on Mount Wilson. In the succeeding decade, three more were built. Indeed, it was a building boom unrivaled in astronomical history since the 1878 to 1897 period that witnessed the construction of eight of the biggest refractors, including the three largest ever built.

The 1969-1979 period included two Soviet telescopes, the Zelenchukskaya monster and a 102-inch telescope at **Byurakan Astrophysical Observatory** in Armenia. The United States built two in Arizona, including the 148-inch **Nicholas Mayall Telescope** at the **Kitt Peak National Observatory**, which opened in 1973, and the **Multiple Mirror Telescope** at the **Whipple Observatory** on Arizona's 550-foot Mount Hopkins, which opened in 1979. The latter, with its distinctive *six* 72-inch mirrors, provides the light-gathering power of a 176-inch telescope. This makes the newest of the great telescopes the world's third largest, and Palomar's only true rival.

Three more important telescopes were opened in the Southern Hemisphere during the same period. These were the first since the refractors built in South Africa in the 1920s. The first was the 152-inch **Anglo-Australian Telescope** atop 3820-foot Siding Spring Mountain in Australia, which opened in 1974. It was in turn followed by three cooperative ventures that opened in Chile in 1976 and 1977. These were the 140-inch **European Southern Observatory** at Cerro La Silla, the 101-inch **Irenee du Pont Telescope** at Cerro Las Campanas and the 156-inch **Inter-American Observatory** at Cerro Tololo.

Perhaps the most important astronomical site to develop in the 1970s, however, was on the island of Hawaii in the mid-Pacific. In 1979, three important observatories were built under the auspices of the University of Hawaii's Institute for Astronomy on top of Mauna Kea, a dormant volcanic peak. They are important primarily for being at higher elevations than any other observatories in the world. Indeed, they are more than twice as high as Mount Palomar and nearly twice as high as Cerro Tololo. These telescopes were the 117-inch **NASA Infrared Telescope** at the 13,650-foot level, the 140-inch **Canada-France-Hawaii Telescope** at the 13,720-foot elevation, and the 148-inch **United Kingdom Infrared Telescope (UKIRT)** at 13,780 feet.

While Hawaii is typically known for its warm beaches, the top of Mauna Kea is often subject to severe blizzards and heavy snow. However, the altitude, accompanied by the distance from major population centers, makes Mauna Kea one of the best observation centers in the world.

The 1980s saw the development of a series of major new observatories in Spain—two of them on La Palma in the Canary Islands—that were inaugurated between 1984 and 1987. These were the 100-inch **Isaac Newton Telescope**, moved from England to the **Observatorio Roque de los Muchachos** at La Palma, the 137-inch **German-Spanish Astronomical Center** at Calar Alto and the 164-inch **William Herschel Telescope** built in England for the Observatorio Roque de los Muchachos, which became the world's fourth largest.

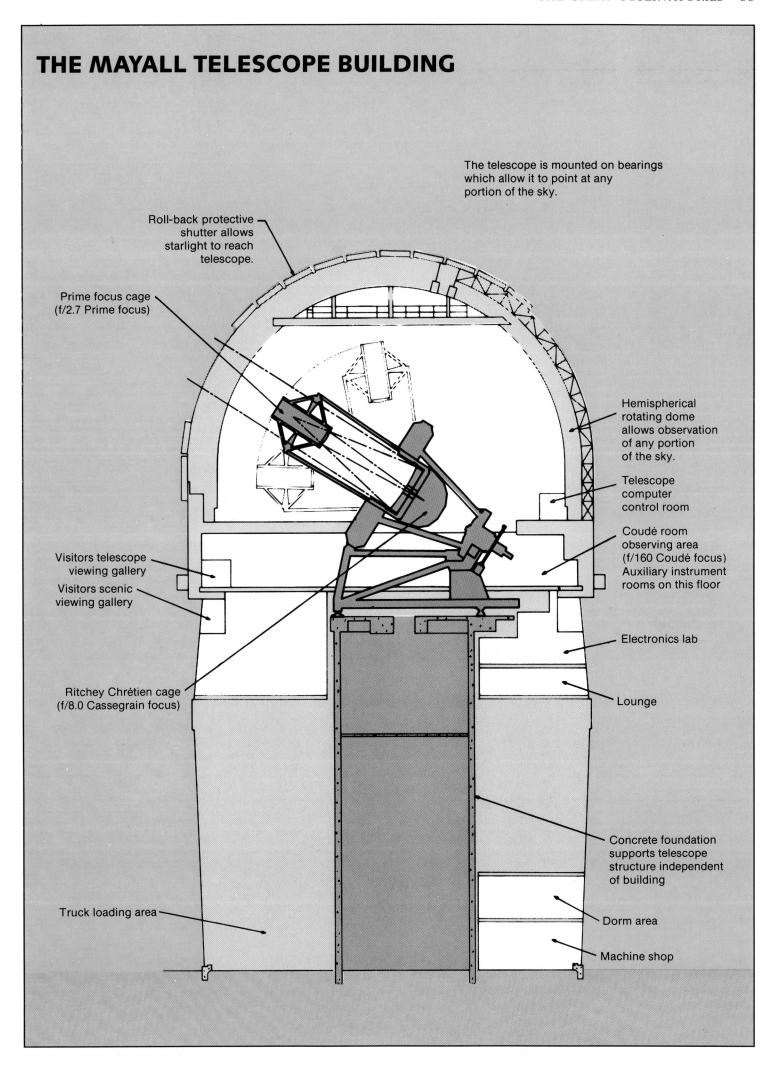

The next important telescope to go on line will be the 390-inch WM Keck Observatory being developed by the University of California at the peak of Mauna Kea's unrivaled vantage point. Several other large telescopes, exceeding 200 inches, are also being planned for Arizona and Chile. The largest on the drawing boards in the early 1990s is the 585-inch American **National New Technology Telescope (NNTT)** to be administered by the United States government's **National Optical Astronomy Observatories**. This telescope, more than double the size of Hale's Palomar instrument, would provide a glimpse into the mysteries of deep space that were still mere dreams in Hale's era.

Since Hale's time, the goal in optical astronomy has been to get higher and higher, into the thinnest reaches of the atmosphere. This has been the appeal of Mauna Kea. Ultimately, however, the future of optical Astronomy may rest with the successors to **NASA's Hubble Space Telescope**, which can observe the universe from high above all the hazy effects of the Earth's atmosphere.

In the 1930s, it was discovered that radio waves were reaching the Earth from deep space and soon astronomers were began to study radiation from other than the narrow slice of the spectrum that was visible optically. Thus was born the science of **Radio Astronomy**.

It was **Karl Jansky (1905-1950)** who first discovered extraterrestrial radio 'interference.' Although he did some pioneering work in the field, Grote Reber at Chicago built the first radio telescope for practical observation in 1937. Technology developed during World War II—particularly with respect to radar—really boosted the development of the new science.

The first great radio 'telescopes' (not really telescopes, but dish antennas) was completed at **Jodrell Bank** in England in 1957. Measuring 250 feet, it was the state-of-the-art until the completion of the American 300-foot transit radio telescope at Green Bank, West Virginia in 1962. A year later, **Arecibo Radio Observatory** in Puerto Rico, with its 1000-foot dish, was opened. It remains the largest, fixed, single-dish radio telescope in the world. Built by Cornell University under contract to the Department of Defense, it was turned over to the National Science Foundation in 1969 as part of the **National Astronomy & Ionosphere Center**.

The largest *steerable* radio telescope in the world (328 feet) was completed in 1972 by the **Max Planck Institute for Radio Astronomy** and is located at Effelsburg in Germany.

The most important new radio observatory in the world is the US National Science Foundation's **Very Large Array** at Socorro, New Mexico, which began operation in 1980. It is not a single dish, but an array of 27 parabolic dishes, each 82 feet across, arranged in a Y-shaped pattern over an area 13 miles by 12 miles. Because of the integration of 27 dishes, the Very Large Array has the effective capability of an antenna 16.75 miles across, or *88 times* that of the Arecibo Observatory!

The Very Large Array has been used to make more detailed radio atlases of the universe than were ever thought practical. So capable is the Very Large Array, that it easily tuned in to the one-watt signal broadcast from the Pioneer 11 spacecraft more than four billion miles in space.

The Very Large Array also is devoting time to NASA's **Search for Extra-Terrestrial Intelligence (SETI)** project, which was begun in 1960 to search for radio signals that could be from intelligent creatures, to do radio astronomy mapping of the sky and study manmade radio interference that could affect space tracking, data acquisition, and communication. Beginning on 12 October 1992, on the 500th anniversary of Columbus' discovery of the New World, the Very Large Array's unprecedented observational and computing power was brought to bear to answer the ultimate question: Are we alone?

Left above: **The Very Large Array radio observatory in Socorro, New Mexico.** *Left below:* **One of the dish antennas at the radio observatory in Owens Valley, CA.** *Right:* **Karl Jansky in front of his antenna nicknamed the 'merry-go-round.' It was mounted on four Model-T wheels and rotated around a circular brick track once every 20 minutes.**

Stellar Astronomy

With the advent of truly powerful telescopes in the late nineteenth century, the science of **Stellar Astronomy** made enormous progress. The systematic measurements of stellar brightness, initiated by the Herschels, was continued in 1879 by **BA Gould (1824-1896)** with his *Uranometria Argentina*, which gave estimated magnitudes of 8000 stars visible from Cordoba, Argentina. In 1884, **EC Pickering (1846-1919)** of Harvard College Observatory published a catalogue of 4260 stars constructed from nearly 95,000 observations made with a meridian photometer between 1879 and 1882. The **meridian photometer** consists of a pair of fixed horizontal telescopes, each with a movable mirror before the object glass. One mirror directs the light of the Pole Star to the eye of the observer, and the other directs the light of the star to be measured. By means of a polarizing device, the light of the brighter star is dimmed until it is apparently equal to the other. A magnitude of 2.1 is assumed for the Pole Star, and the light absorbed by the polarizing system being known, the magnitude of the star to be measured can be obtained.

In 1885, **C Pritchard (1808-1893)** of Oxford University issued his *Uranometria Nova Oxoniensis* with magnitudes,

Below: **The Trifid Nebula in Sagittarius, M20, NGC 6514.**
Opposite: **The Pleiades in Taurus, M45, NGC 1432.**

made by a wedge photometer, of all naked eye stars from the North Pole to 10 degrees south of the celestial equator, numbering 2784. In the **wedge photometer**, an 'artificial star' is formed by a suitable combination of lenses and diaphragms, and brought, by reflection, into the field of view of the telescope close to the image of the star to be measured. By means of a graduated wedge of neutral tinted glass moved across the artificial star's image until equality of the two is observed, the desired value of magnitude is found.

The agreement between these two catalogues was fairly good. Between 1889 and 1891, the Harvard photometry was extended to the south pole of the sky, and between 1891 and 1898, it was enlarged to cover all stars brighter than 7.5 magnitude from the North Pole to 40 degrees below the celestial equator. Another catalogue, the *Potsdamer Photometrische Durchmusterung*, issued in parts from 1894 for the next 13 years, gave the magnitudes of more than 14,000 stars from the North Pole to the celestial equator, down to 7.5 magnitude.

In these catalogues the ratio between the brightness of successive magnitudes was taken as 2.512 to 1, the so-called **Pogson's Ratio**, which had been proposed in 1856 by **N Pogson (1829-1891)**, and based on J Herschel's observation in 1830 that an average first magnitude star is 100 times brighter than one of the sixth magnitude. Before the general adoption of this ratio, the relative brightness of stars was very vaguely defined. Herschel, Struve, and Argelander used scales which differed widely from each other.

The measurements of parallax by Bessel, Henderson, Struve and others during the first part of the century, were not followed by many results of high quality for some time. In 1840, the three values referred to earlier were known (for 61 Cygni, Alpha Centauri and Vega). By 1880, this had increased only to 20, and in 1900 to 60, the most reliable being those found with the heliometer, although photography was introduced in 1886 with promising results by C Pritchard at Oxford.

Near the middle of the nineteenth century, spectroscopy became common in Stellar Astronomy. The first to apply the modern spectroscope to the stars, apart from Fraunhofer's early work, was the same astronomer who had done so for comets—G Donati of Florence. With his instrumented equipment, he was unable to do more than note the positions of a few of the prominent lines. However, the real founders of stellar spectroscopy were **A Secchi (1818-1878)** and W Hug-

gins. Secchi completed a general survey for classification purposes. Huggins did a detailed study of particular spectra.

Between 1863 and 1867, Secchi's classification into four types was of very great utility, until superseded by a system originating at Harvard. He applied his system to more than 4000 stars of a catalogue. The first type of star included bluish-white or white stars, like Sirius and Vega, and had prominent lines of hydrogen. The second type of star included yellowish stars like the Sun, Capella and Arcturus, which showed many lines of metals. The third, or red stars, such as Antares and Betelgeuse, had spectra with strong dark bands. The fourth type of star was a less numerous class of red stars with a different kind of absorption bands. A fifth class of stars, showing bright lines or bands, as well as dark lines on a continuous spectrum, was added in 1867 by Wolf and Rayet of Paris the 'Wolf-Rayet' stars.

In 1864, Huggins first observed the spectrum of a diffuse nebula, which he noted consisted of several bright lines. During the next four years, he examined the spectra of 70 nebulae, finding bright lines for more than 20, the most notable being the Orion nebula. A large proportion of the nebulae gave a continuous spectrum, indicating that they were probably unresolved star clusters. His findings also showed the gaseous nature of a considerable proportion of nebulae. Initially, only hydrogen could be identified as a constituent; one of the lines first observed was due to this gas.

Identification of lines of known elements—such as hydrogen, iron, sodium and calcium—first noted in stellar spectra by Huggins in 1864, was followed by numerous other identifications by spectroscopists. In the same year, Huggins advocated the principle that the color of a star depends on the absorption of its light by its surrounding atmospheric vapors. This theory became more popular than the more correct theory advanced by Zollner in 1865 that color depends chiefly on temperature, so that yellow and red stars were in progressive stages of cooling. In 1874, **HC Vogel (1842-1907)** adopted this theory, stating that Vega and Sirius were the youngest type of stars, and yellow and red stars—like Capella or Betelgeuse—were older and cooler. Vogel's system of classification resembled Secchi's in a general way, and in 1885 he revised it to include a newly recognized type of bluish white star called the 'helium' stars because of the helium lines in their spectra.

In 1887, J Norman Lockyer advanced his **Meteoritic Hypothesis**, the basic proposition of which is that 'all self-luminous bodies in the celestial spaces are composed either of swarms of meteorites or of masses of meteoric vapor produced by heat.' The common origin thus suggested was supposed to result in a progressive evolutionary course whereby nebulae and stars were disposed along a temperature curve, rising from nebulae and bright-line stars through red stars of Secchi's third type, then second type yellow stars, to the Sirian first type. From this stage the descent in temperature was through second type (solar) stars to red stars of the fourth type, and then to extinction. However, later spectroscopic work showed that certain supposed coincidences of spectral lines, on which the meteoritic part of the hypothesis rested, were not in fact exact enough.

In 1890, the *Draper Catalogue of Stellar Spectra* was published by Harvard College Observatory. For this catalogue, the spectra were obtained 'in bulk' by use of a **prismatic camera**, which consists of a photographic telescope with a large prism of a rather small angle, an objective prism, in front of its object glass. The spectra appear as narrow streaks on a plate placed in the focal plane of the telescope. This catalogue included particulars and classification for more than 19,000 stars down to the eighth magnitude. The classification scheme, which was destined, with modifications, to take the place of all others, was divided into types designated by letters of the alphabet. The chief divisions used are O, B, A, F, G, K, M and N. Class O are the Wolf-Rayet stars; B and A roughly correspond to Secchi's first type; G and K are broadly Secchi's second type; M is Secchi's third type; and N is Secchi's fourth type.

In 1868, Huggins applied spectrum analysis to the problem of the motions of the stars by measuring the position of the lines to find if there was any displacement that might be attributed to motion by Doppler's principle. He used the brightest star, Sirius, for this purpose, and found distinct evidence of a motion of recession, later verified by Vogel. The next year, Huggins determined the line-of-sight movements of 30 stars, and it may be said that the study of stellar radial velocities dates from these investigations. On radial or—which is perhaps a better name—line-of-sight velocities, plus proper motions and a knowledge of distances to convert these proper motions into actual velocities, depends the solution of the speed and direction of movement in space of any heavenly body extraneous to our Solar System. Photography was later applied to line-of-sight velocity determination with greatly increased accuracy as visual observation and measurement could not generally be so accurate with the difficulties to be overcome.

The first to use photography for this purpose was Vogel in 1887, and he was followed by others, notably Keeler, Frost and Campbell. The data Vogel accumulated later enabled Campbell to find, from the line-of-sight velocities of 280 stars, the values for the speed and direction of the Sun's movement in space. In 1890, Keeler, using the powerful telescope at Lick Observatory, was able to visually measure the velocities in the line-of-sight of some bright line nebulae, which Huggins had not been able to do in 1874.

The first to photograph a star's spectrum was H Draper in 1872. Four years later, Huggins used the dry plate, for the first time in such work, for the same purpose, on the same star, Vega. The first photographs of the spectrum of a nova were made on Nova Aurigas 1891. In regard to nebulae, Huggins was successful in photographing the bright-line spectrum of the Orion nebula in 1882, and **J Scheiner (1858-1913)** in January 1899 photographed the spectrum of the great Andromeda nebula, finding it to look like a cluster of solar stars.

Notable milestones in general astronomical photography were: the Orion nebula, photographed by Common using a 3-foot aperture reflector in 1883; star fields by the Henry brothers in 1885; discovery of a nebula in the Pleiades by the Henry brothers, using a 13-inch aperture refractor, in 1885; the Andromeda nebula, showing some traces of spiral structure by Roberts, in 1886; Pleiades nebulosities by Roberts, using a 20-inch aperture reflector, in 1886; Orion nebula connections, covering a large part of the constellation, by WH Pickering, using a small aperture star camera, in 1889; and star clouds of the Milky Way by Barnard, using a 6-inch aperture star camera with a portrait lens, in 1889.

Although certain stars had been seen to be double or multiple, notably Zeta Ursae Majoris (one of the stars in the Big Dipper) by Riccioli in the seventeenth century, Theta Orionis (multiple) by Huygens in 1656, and Gamma Arietis in 1664 by Hooke, Herschel's first survey of the sky in 1779 may be said to mark the real beginning of **double star Astronomy**. In the same year, **C Mayer (1709-1783)** published a small book in which he speculated on the probable existence of binary systems and gave a list of 89 pairs. After Herschel's time and the end of the nineteenth century the following outstanding workers in the subject may be specially mentioned: FGW Struve, WR Dawes, O Struve, **Baron E Dembowski (1815-1881)** and **SW Burnham (1838-1921)**. Dawes, an amateur using telescopes ranging from four to eight inches, made measurements of hundreds of double stars, which he summarized and published in 1867. Struve published a catalogue of 514 stars, of which a very large proportion were close pairs. Baron Dembowski, an amateur observing in Italy, measured pairs for more than 30 years, using five and seven-inch refractors. Burnham, beginning as an amateur in Chicago with a six-inch refractor, probably used, as a professional, a greater range of large telescopes than any one else (nine instruments at various observatories up to the 40-inch Yerkes refractor), discovering 1290 new doubles from 1871 to 1899. After his transfer from the Lick Observatory to Yerkes, his successors at Lick, **WT Hussey (1862-1926)** and RG Aitken discovered hundreds of new pairs before the end of the century. A catalogue for the southern skies was published in 1899 by **RTA Innes (1861-1932)** containing 2140 pairs, more than 300 of which he had discovered himself. In 1897, TJJ See found nearly 500 new southern pairs with the Lowell 24-inch refractor.

Meanwhile, Bessel had predicted from the disturbed proper motions of Sirius and of Procyon, that each of these stars would be found to have a faint stellar companion, revolving round them in periods, in both cases, of about half a century. These were telescopically discovered. The former was by AG Clark, the famous optician, in 1862, when testing a new 18-inch refractor, and the companion to Procyon in 1896 by Schaeberle, using the 36-inch refractor at Lick.

In the photography of double stars, as far back as 1857, GP Bond of Harvard obtained a collodion plate giving measurable images of Zeta Ursae Majoris, a rather wide double. EC Pickering later made a few attempts, and FA Gould obtained some results between 1870 and 1882 at Cordoba. In 1886, the Henry brothers at Paris secured successful photographs of several pairs ranging from about 21.33 to over five seconds separation, and in 1894 and 1897 good results were obtained at Greenwich with 13-inch and 27-inch refractors, yielding measurable images down to 1.25 seconds apart. Photography, although providing measurements of great accuracy for such pairs, did not seem to be yet applicable for very close systems a second of arc or less apart.

A new type of binary star was discovered in 1889 by EC Pickering at Harvard. On plates taken for the *Draper Catalogue of Stellar Spectra,* it was noted that a prominent line in the spectrum of the brighter component of Zeta Ursae Majoris (the first star to be seen visually as a pair) occasionally appeared doubled. This was interpreted to mean that the star had two components revolving around one another, and that their motions to and from the Earth—line-of-sight velocities—are changing regularly, one star receding while the other is approaching, and their spectra therefore had lines which shifted with reference to one another. The period of revolution was found to be 20.1 days, much shorter than the several years of the shortest period visual binary. In 1880, Pickering had shown from the shape of the light curve of the variable Algol that the suggestion of **J Goodricke (1764-1786)** as to the cause of its variation was correct, ie, that an eclipse of a brighter star by a fainter component revolving around it in the period of light change was responsible. Pickering was also able to calculate their relative dimensions and distance apart. In the same year (1889) as the spectroscopic discovery of the binary nature of Zeta Ursae Majoris, Vogel of Potsdam found that Algol also showed

Below: **A diagram of the quadruple star system in the constellation Leo. The M-class companion binary dwarf stars orbit the K-class contact binary star XY LEO once every 20 years.**

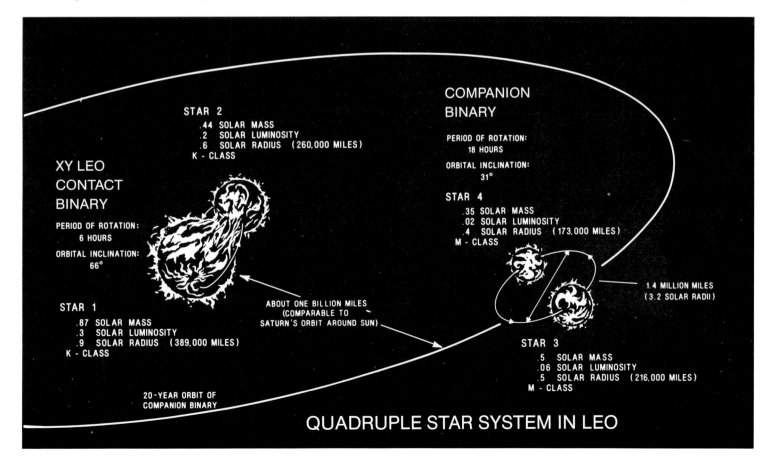

displacement in the lines of its spectrum, thus confirming Pickering's theoretical investigation and Goodricke's previous surmise. Only the spectrum of the brighter star is visible in this case, but the shifts of the lines have a period the same as that of the light variation, about 2.85 days.

A number of stars were soon discovered for which, like Zeta Ursae Majoris, there is no eclipse, the plane of revolution of the component stars not being in the line of sight. By the end of the century, 13 of these **spectroscopic binaries** were known. Few if any of such systems are ever likely to be seen as a visual double, as the components are too close for separation by the largest telescope. The power of separation of a telescope is proportional to aperture. A four-inch aperture can separate a pair about a second of arc apart; a 40-inch aperture separates two stars a tenth of a second from each other. The separation of these spectroscopic binaries will be usually much less than this lower figure. In the case of Algol, this is probably between one 1/400th and 1/500th of a second of arc, requiring an aperture of about 170 feet. One exception is perhaps the bright star Capella, discovered to be a spectroscopic binary in 1899 by Campbell and Newall independently. In this case, it has been claimed that elongation of the image in the correct direction has been seen with the 28-inch Greenwich refractor.

Apart from what appears to be a strong probability that the eclipsing variable Beta Persei was known to the Arabians to change in light (they named it Algol or The Demon), the first known variable, Mira Ceti, was discovered by D Fabricius in 1597. The early history of Mira is briefly as follows. D Fabricius saw it as third magnitude star in April 1596 and noted that it disappeared the following year. Bayer gave it the Greek letter Omicron of the constellation Cetus, but did not seem to have connected it with Fabricius' observations. No further notices of it appear to have been recorded until 1638 when Holward found it third magnitude, invisible in the summer of 1639, and again visible in October 1639. From 1648 to 1662 Hevelius carefully observed its changes, and since that time Mira seems to have been followed by someone until the present day. In 1669, **G Montanari (1632-1687)** discovered (or rediscovered?) the variability of Algol, R Hydrae was found in 1670 by the same observer, and Chi Cygni by Kirch in 1686. No more were found until 1782, when Goodricke detected the variability of Beta Lyrae and Delta Cephei, and another seven were added before 1800, making 13 in all.

An astronomer who greatly contributed to the study of variable stars was Argelander, who introduced a procedure of 'step' comparison between variable and nonvariable neighboring stars of the same brightness (a method of intercomparison of magnitude employed earlier by W Herschel). He also advocated cooperation in observation, especially addressed to amateurs.

In 1854, Pogson published a list of 53 variables, and Schonfeld, Argelander's successor, published a list of 113 in 1865 and 165 in 1875, while in 1884 and 1888 JE Gore's catalogues contained 190 and 243, respectively. Soon afterward, EC Pickering's introduction of comparison by photographs taken at

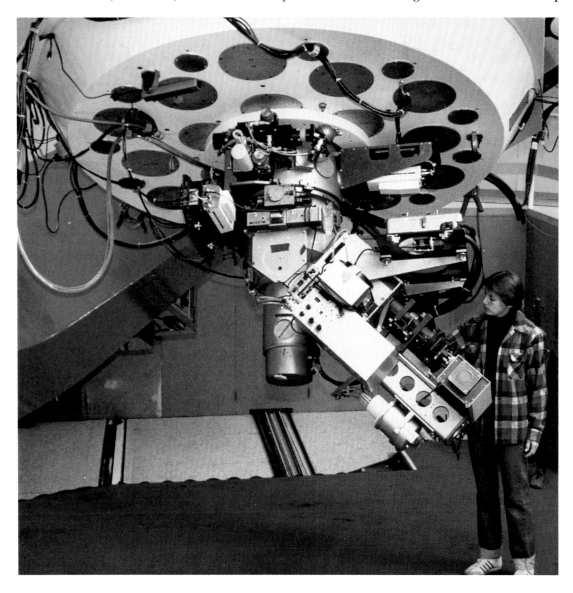

Left: **Astronomers Drs Catherine Pilachowski and Carol A Christian examine the 83-inch telescope at Kitt Peak.**
Right above: **The Eagle Nebula in Serpens, M16, NGC 6611.**
Right below: **The Great Nebula in Orion, M42, NGC 1976.**

OBSERVING NOVAE

In early Chinese accounts, **Novae** are referred to as 'K'o-hsing' or 'guest stars,' distinguished from the moving comets by their stationary position with reference to surrounding stars, and by the records of their appearance and disappearance. Their brightness compared with the planets or fixed stars, and a progressive account of their fading are sometimes given. There are two types of Novae: ordinary, which are about 50,000 times as bright as the Sun and **Supernovae**, which may outshine the Sun about 10 million to 100 million times. In the 1930s, investigations based on Chinese observations of a Supernova seen in 1054 AD showed that a well-known nebula in the constellation Taurus remained after an outburst seen nearly a thousand years ago.

The details of the ancient Chinese and Japanese observations are of great interest, and the investigation has led to the review of similar ancient records which might help in the establishment of a connection between Novae of the past with some of the nebulous objects in the sky.

different dates for detection of variables greatly increased the rate of discovery, and his adoption of instrumental photometry for variables in 1896 would do much to encourage precise knowledge in the subject.

In 1880, Pickering had proposed a system of classification of variables, which formed the basis of all subsequent schemes. It may be briefly described as follows: the first class, novae or temporary stars; the second class, long period variables, like Mira Ceti; the third, irregular variables, like Betelgeuse; the fourth, short period variables, such as Beta Lyrae or Delta Cephei; and the fifth, eclipsing variables such as Algol. Later on, Beta Lyrae was found to be an eclipsing variable, and in 1895 S Bailey of Harvard detected a type of very short period— less than a day—common in certain of the globular clusters of stars. It is interesting to note in passing, that the first six variables, known by 1782, include examples of the three main classes, long-period (Mira Ceti, R Hydrae, Chi Cygni), short period (Delta Cephei), and eclipsing (Algol, Beta Lyrae).

Observations of three short period variables, Chi Cephei, Xi Geminorum, and Eta Aquilae by Belopolsky in 1894 and 1895, suggested that these stars, which show rapid, regular fluctuation in displacements of lines in their spectra, are accompanied by companions, revolving in the same period as their light variation, but without eclipse of their primaries.

Later research showed the impossibility of this, and a different explanation was given, attributing pulsation to their bodies.

In the seventeenth century, three novae were recorded, two in Europe in 1604 and 1670, and a Chinese observation in 1609. Another star of perhaps the nova type, known as P Cygni, was seen in 1600 as a third magnitude star by W Janson. It has been only about a quarter as bright for many years since. In the eighteenth century, there were none. In fact, for 178 years until the discovery of a nova in Ophiuchus in 1848, of only the fifth magnitude, by **JR Hind (1823-1895)**, there were no novae observed, although in the long interval the skies were certainly more carefully examined than at any previous time, with the advantage of telescopes and improved star catalogues.

The new star of 1866 in Corona Borealis found by J Birmingham, an Irish amateur, was therefore of particular interest to astronomers. It reached the second magnitude, and some time after its discovery Huggins observed bright lines in its spectrum, particularly of hydrogen. The next nova was detected by Schmidt in 1876 in the constellation Cygnus. It was of the third magnitude and showed a spectrum of bright lines of hydrogen and helium on a continuous background with strong absorption lines.

In 1885 a nova, which reached seventh magnitude, was seen in the Great Andromeda Nebula in a position where no star as bright as the fifteenth magnitude had previously existed, and in 1860 another seventh magnitude nova appeared near the center of the globular cluster M80 in the constellation Scorpio. In 1892, one which attained fourth magnitude was found in Auriga, and in February 1901, another in Perseus which at its brightest outshone Capella, both by Anderson of Edinburgh, an amateur. These two were closely studied by means of the spectroscope and showed the usual type of nova spectrum.

A number of fainter novae were found from 1893 onwards by a new method. This was from their spectra as photographed with a prismatic camera, but such discoveries were fortuitous rather than systematic and did not, except by accident, show the spectrum when the star was at its brightest.

Nova Persei was the brightest since the Nova of 1604. A large nebula extending southeast from it was photographed by Wolf with a 16-inch star camera. Further photographs taken at Yerkes and Lick observatories showed a movement of condensations in the nebula away from the star. This was explained by Kepteyn as the gradual illumination, by the light of the nova, of a nebulous cloud already existing in the surrounding space.

Several attempts to account for the phenomena of a nova were put forward about this time. Of these, two may be mentioned. **WHS Monck (1839-1915)**, and **H Seeliger (1849-1924)** independently, in 1892 suggested that a temporary star appears when a dark body or bodies are made luminous by friction while passing through a cloud of nebulous

Left: **Before *(far left)* and after *(left)* photographs of Supernova 1987A in the Large Magellanic Cloud. Taken in 1969 and 1987, respectively, the later image shows the bright supernova detected in this nearby galaxy.** *Right:* **Andromeda Galaxy, M31, taken by the 200-inch telescope at the Palomar Observatory.** *Overleaf:* **The Lagoon Nebula in Sagittarius M8, NGC 6523.**

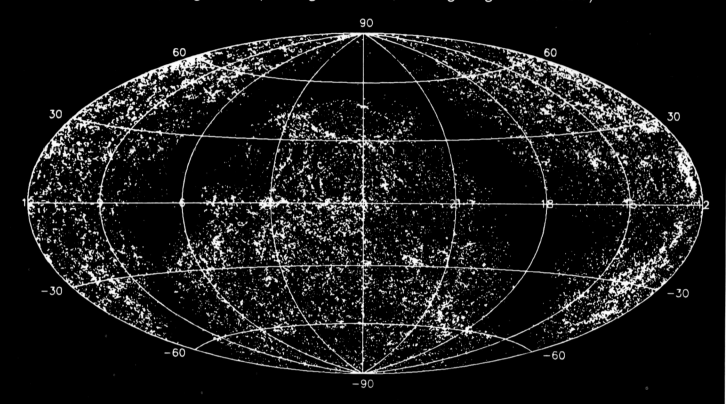

NA Sharp *PASP* 98 740 (1986)
34,729 galaxies (limiting size~0.3', limiting magnitude~15.5)

Right Ascension & Declination

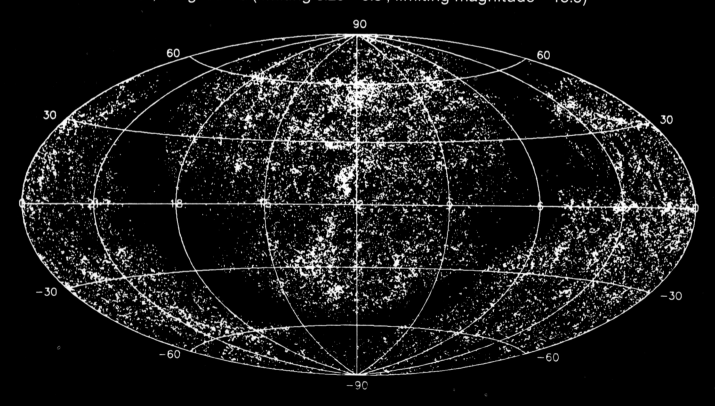

NA Sharp *PASP* 98 740 (1986)
34,729 galaxies (limiting size~0.3', limiting magnitude~15.5)

Right Ascension & Declination

matter. This, they thought, would give the spectrum observed, ie, continuous from the incandescent body, and bright lines from the gaseous nebula. The other suggestion was of a collision between two stars, of a grazing nature, forming a gaseous 'third body' composed of the coalesced detached parts of the colliding stars. This was proposed by **AW Bickerton (1842-1927)** in 1879. Neither hypothesis seems to have survived later criticism.

In 1869, RA Proctor found that five of the seven stars of the Big Dipper have a movement across the sky in the same direction and with the same angular velocity (about 1100 miles in 18,000 years), this motion being clearly not a reflex of a movement of the Solar System in space. He concluded that they form a connected moving group, and he also noted community of motion among certain stars in Taurus, years afterwards studied as the Taurus Moving Cluster.

Huggins examined the five stars of the Big Dipper with the spectroscope and found them to be moving in the line of sight away from us at the same rate of speed, thus putting the question of their connection beyond doubt. This was the first intimation of such group motion. Nevertheless the general opinion at the end of the nineteenth century was that the stars move generally at random, although in 1895 Kobold believed he had found evidence of some general motion in opposite directions in the plane of the Milky Way.

The employment of photography for nebular research has already been mentioned. H Draper, Common, Wolf, **I Roberts (1829-1903)**, Barnard and Keeler were all pioneers in this work which was extended to discovery as well as to detailed study. The catalogues of nebulae and clusters of the Herschels were used as the basis for the *New General Catalogue* of **JL Dreyer (1852-1926)** published in 1888. It contains the particulars for 7840 nebulae and clusters and was supplemented in 1894 by an *Index Catalogue* with 1529 additional objects. Using the 36-inch Crossley reflector at Lick Observatory, Keeler photographed thousands of nebulae. He estimated that 120,000 new nebulae were within its reach, and that probably all are of the spiral type. These large numbers have, however, proved to be underestimates.

In 1894, Wolf discovered photographically in a nebulous region three dark markings in the Milky Way about 1.5 degrees east of Gamma Aquilae. Further east, Barnard found other curiously shaped dark markings, and in 1895 he photographed an extensive nebulosity in Ophiuchus in which there are long dark lanes devoid of nebulosity and with practically no stars. These and many other markings of the kind were taken to be of a similar nature to the 'Coal Sack' in the southern Milky Way and were believed by most at the time to be openings through the star clouds.

Prior to the discovery by Huggins in 1864 of the gaseous nature of diffuse nebulae, there had been a tendency to assume that all the nebulae are clusters of stars, not resolvable into individual stars owing to distance. But many, departing from the 'external universe' idea of Herschel, held that all types of nebulae would be found to be members of the Milky Way system. As pointed out by Herbert Spencer and by RA Proctor, this was apparently supported by the facts of their peculiar distribution—much more numerous at the Milky Way polar regions than towards the Galaxy. However, the view was not held rigidly by all. Even Proctor wrote in 1869 that it is 'not improbable that the spiral nebulae are galaxies resembling our own.' The observation of continuous spectra in the Andromeda nebulae and in others situated away from the Milky Way area, was seen to be consistent with this idea.

Left: **These Aitoff equal area projections show the distribution of galaxies in the sky with the *top* image centered at a right ascension of zero hours and the *bottom* centered at 12 hours to compensate for the distortion.** *Above:* **The Milky Way.**

Nevertheless, the majority view was against the 'external universe' hypothesis, especially with what appeared to be an impossibly high luminosity for the 1885 nova in the Andromeda nebula at the distance entailed.

Various views of the shape and extension of our Galactic system were expressed during the nineteenth century. J Herschel seemed inclined to the idea of a ring formation, sized roughly 4000 light years in diameter, although he never expressly disassociates himself from his father's disc theory in any of his writings. Proctor advocated that it consists of a stream of a twisted shape, composed of comparatively small stars, but gave no estimate of dimensions.

JE Gore considered it to be a ring-shaped cluster reduced to a nebulous appearance in the Milky Way through great distance, and gave various estimates of size from 6000 to 20,000 light years. H Seeliger thought that it was a disc formation 23,000 light years in diameter, while **S Newcomb (1835-1909)** believed the size to be at least 6000 light years. C Easton of Rotterdam, thought that our Solar System was a double-armed spiral, with the Sun near the center of the system in all cases referred to except the last. In Easton's view the center of mass of the system and the nucleus of the spiral were in the star clouds of the constellation Cygnus, with the Sun situated in a comparatively vacant space near the center of the volume occupied by the spiral and its arms.

Solar Astronomy

Although omnipresent in the sky overhead since before the dawn of time, it was not until the nineteenth century that astronomers began to probe the mysteries of the nature of the Sun. Observations of solar eclipses were used during that period chiefly for corrections to the movements of the Sun and Moon, a check on which was provided by the times of apparent contact of the limbs of the two bodies. After the eclipse of 8 July 1842, which was observed in central and southern Europe by the leading astronomers of the day, the appearances surrounding the eclipsed Sun began to be more closely noted and studied. These were the **corona**, its luminous surrounding halo of light, and the colored **prominences** seen on the edge of the disc. Both of these phenomena had been noted before and had sometimes been considered to be appendages of the Sun, but about as often had been believed to be lunar features. After the 1842 eclipse, study of all total solar eclipses was systematically pursued with a view to a solution. The question of their nature, whether self-luminous solar appendages or lunar clouds lit up by the Sun, was not yet decisively answered. In the eclipses between 1842 and 1851 the process of discovery of the nature of these appearances was continued, but the Wilson-Herschel Theory of the Sun's constitution does not appear to have been affected as yet.

The telescopic study of sunspots at this time resulted in one epoch-making discovery. **Heinrich Schwabe (1789-1875)** of Dessau, began in 1826 to observe the Sun on every clear day with a small telescope, hoping to discover an intra-Mercurial planet. Every day for more than 40 years, he counted the number of spots, and by 1843 after years of patient work, he noted a recurring periodicity of about 10 years in the daily numbers.

In his tabulated figures for 1826 to 1843, the hint of periodicity was strongest in the numbers of days, for a given year, showing no spots. For example, in 1828 there were no such days, five years later in 1833 there were 139, five years after that in 1838 there were again none, to be followed in 1843 by 149. Also, the recorded numbers of no spot days each year between these dates varied progressively. By 1851, Schwabe had confirmed this and he then published his results, describing the 10-year cycle.

Examination of previous records of sunspots and later observations increased Schwabe's estimate to a period of 11.1 years, but this is correct only on an average. Intervals from seven years (between 1830 and 1837) to 16 years (between 1788 and 1804) were noted. From Chinese records between 188 AD and 1638, during which time 95 observations of spots visible to the naked eye were found, an average period of 11 years was obtained. It may be noted, however, that according to the researches of Sporer and Mounder, it appears that there was a period of great scarcity of spots between about 1643 and 1715. Study of the varying thicknesses of the growth rings in giant sequoia trees of California by AE Douglass also shows what is evidently the effect of this long quiet period, and a contemporary scarcity of auroras has been noted. The first to suggest periodicity of sunspots, but without any idea of the time interval, was the Danish astronomer **Christian Horrebow (?-1776)**.

A similar period for the Earth's magnetic variations was first noted by Lamont at Munich in 1851, and confirmed independently by several scientists in the following year, the probable connection with the Sun being quickly deduced.

The granulated structure of the Sun's luminous surface, described by a famous astronomer as 'like a plate of rice soup,'

Below: **A spectacular solar flare spanning 367,000 miles.** *Opposite:* **A dramatic, annular eclipse of the Sun.**

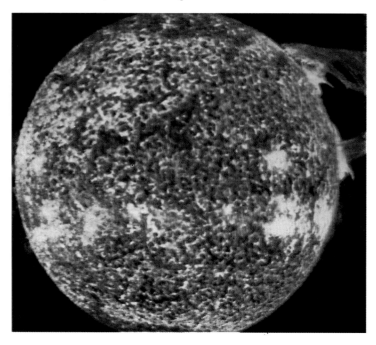

was first recorded by Short, an optician, in 1748. Herschel also detected it, and refers to it in papers printed in 1795 and 1801. Dawes studied it closely in 1830, and Schwabe referred to it in 1831. The study of individual spot structure received special attention from Sir John Herschel at the Cape of Good Hope in 1836 and 1837.

The first determination of a scientific character of the strength of the Sun's radiation were made in 1837 by Sir John Herschel and, a few months later, by the French physicist Pouillet, with nearly identical results. Pouillet found that the vertical rays of the Sun on each square centimeter would raise the temperature of 1.76 grams of water one degree centigrade per minute. The modern value of this, the **solar constant**, which varies one or two percent, is 1.94.

In 1891, the **spectroheliograph** was introduced into the repertoire of the solar investigator. Invented independently both by **George Ellery Hale (1868-1938)** and **H Deslandres (1854-1948)**, this instrument could photograph the Sun exclusively by light of any desired wavelength. Its essential feature was introduction of a second slit just in front of the photographic plate, which thus permitted only a nearly monochromatic beam, chosen at will, to affect the plate. If an image of the Sun was formed (by the object-glass of a large telescope) on the first, or ordinary, slit of the spectroscope, this slit (which had to be longer than the diameter of the image) cut a narrow segment out of the solar image. An image of this segment, produced solely by light of the single wavelength which is isolated by the second slit, is formed on the plate. The Sun's image was permitted to drift slowly over the first slit while the plate was moved with the same velocity behind the second slit. Therefore, the image of segment after segment of the disc was recorded on the plate, until a complete photograph of the Sun had been built up in the selected wavelength.

In photographing the prominences, the disc of the Sun was concealed by an opaque screen of proper size, and the second slit adjusted to pass radiation corresponding to one of the strong, bright lines in the prominence spectrum. Regular records of the whole circumference of the Sun could thus be obtained daily at a number of observatories.

In 1854, Hansen, and in 1858, Le Verrier, questioned the accuracy of the then accepted **solar parallax** (8.58 seconds) and distance (then calculated at 95,370,000 miles) of the Sun, maintaining that the Earth was substantially nearer to the Sun. Their conclusion was based on consideration of the apparent monthly variation in the motion of the Sun across the sky, which reflected a real monthly movement of the Earth around its common center of gravity with the Moon, and depended on the ratio of the distances of the Sun and Moon from the Earth. The amount of this variation being known, and the distance of the Moon already accurately ascertained, a value for the Sun's distance was calculated. According to Le Verrier, this was four million miles less than the accepted amount, with a corresponding parallax of over 8-9 seconds. The periodic variation of Earth's magnetic force had first been noted in 1851 by **J Lamont (1805-1879)**, of Munich, a Scotsman naturalized as a German. He calculated that a period of 101 years fit the observations of the varying

These pages: **The Big Dome Telescope at Sacramento Peak, NM analyzes the Sun in several different ways simultaneously. Instruments include a coronagraph and a magnetograph.**

daily range of magnetic declination. In 1852, **E Sabine (1788-1883)**, **R Wolf (1816-1893)**, and **A Gautier (1793-1881)** independently drew attention to the similarity between sunspot variations and all features of Earth's magnetic disturbances, including auroras, which were greatest when sunspots were most frequent, and vice-versa.

Wolf's exhaustive research in 1852 among records in books, journals and proceedings of societies dating back as far as 1610, found that the average sunspot period was 11.1 years. His conclusion was also based on casual observations of important sunspots and groups of sunspots, and systematic surveys by observers, both amateur and experienced, using large and small telescopes. He reduced all of these observations to one footing by fanning a representative number for each month in a formula which allowed arbitrarily for all factors. The essential reliability of his numbers was shown by their fairly close relation to the areas of sunspots measured on photographs and drawings from 1832 onward, a Wolf number being on the average about one-twelfth of the area of sunspots in millionths of the Sun's disc.

A remarkable solar phenomenon, observed on 1 September 1859 by **RC Carrington (1826-1875)**, was an outburst of two patches of intense light, inside the space covered by a group of spots, which lasted five minutes. At the time of its occurrence, there was a great disturbance of the Earth's magnetic forces that was accompanied by auroral displays, and at the exact time of his observation, a magnetic storm was observed at Kew to become intensified. Carrington's observation was confirmed by Hodgson, and, according to Brodie, was repeated in a less intensified form on 1 October 1864. On 15 July 1892, Hale and others observed a similar outburst.

Carrington, timing the returns of spots situated at different latitudes, showed that the Sun's period of rotation varied from about 25 days at the solar equator to 27.5 days at about 50 degrees north and south latitude. He also noted that spots were seldom seen near the Sun's equator, more frequently in the adjoining zones to 35 degree latitudes north or south, and more sparsely again in the higher latitudes. The work of **FWG Sporer (1822-1895)** closely followed Carrington's investigations, with the principal result now known as **Sporer's Law**. This law held that at the beginning of a cycle, the spots appear mostly in higher latitudes, but as the cycle goes on, they tend to appear at lower and lower latitudes until a position near the equator is reached. Actual sunspot cycles overlap, however, with the new cycle beginning before the expiring one has completed its course.

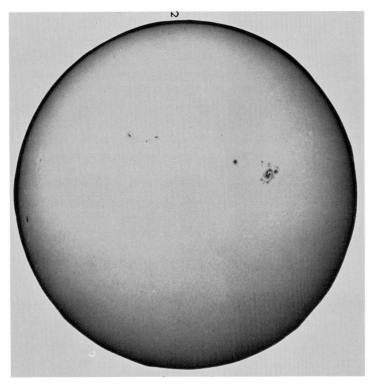

Sporer also showed that, although there was a balance of the numbers of spots between the northern and southern hemispheres of the Sun in the long run, yet in three periods of the Sun's history the southern spots had predominated. He also drew attention to the fact that during the seventeenth century, there was a suspension of the law of the spot cycle and the law of their appearance in latitude for about 70 years, a time when the number of spots was abnormally small.

Wilson's theory that spots were depressions seemed to be confirmed by examination of a series of photographs taken at Kew Observatory from 1858 to 1872. On the other hand, **F Howlett (1820-1907)**, a student of the Sun's surface for 35 years, who made several thousand drawings of spots, published evidence in 1894 that his observations disagreed with Wilson's theory. This controversy faded away when later study of solar phenomena seemed to indicate that the point is perhaps of little significance.

In 1873, a department was opened at Greenwich for the study of the Sun's surface on a daily basis by photography. Not long afterward, arrangements were made to have photographs taken at other observatories connected with Greenwich—one at the Cape of Good Hope and two in India—providing a nearly continuous record of the Sun's surface. Analysis of the records of sunspots and other phenomena, such as faculae—bright streaks or patches on the Sun's surface—by **EW Mounder (1851-1928)** and others, provided further information of great value to the science of solar physics.

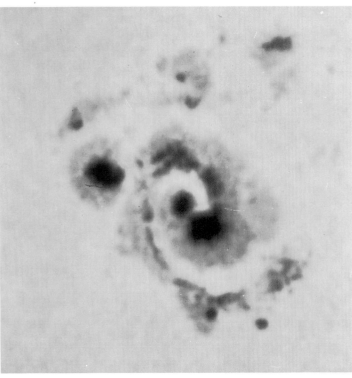

Early observations of the granulated appearance of the Sun's surface have been referred to. In 1861, Nasmyth's announcement that the granulated appearance resulted from the interlacing of willow-leaf shaped forms became the subject of keen discussion. Thirty years later, **PJC Janssen (1824-1907)**, Director of the Meudon Observatory in France,

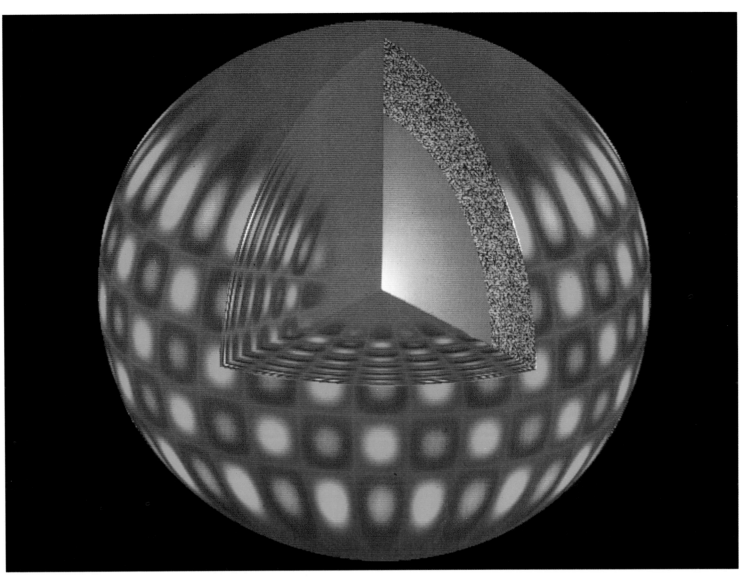

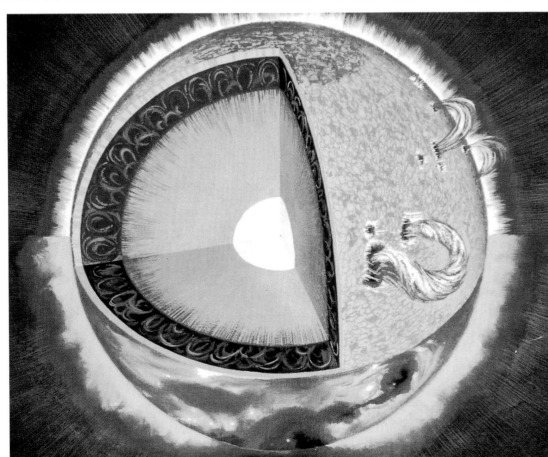

Opposite top and bottom: A white light photo of the Sun, showing sunspots including a rare spiral sunspot. The unprecedented spiral sunspot had a diameter about six times that of the Earth and held its shape for two days before it broke up and changed its form. *Above:* Helioseismologists measure the oscillations of soundwaves of the Sun. This computer representation shows receding regions in red and approaching regions in blue. *Right:* A cutaway view of the Sun showing the core, the convection zone, the photosphere, the chromosphere, a solar prominence and sunspots.

Above: **Erecting a solar monitoring telescope in Antarctica.**
Right above: **A total solar eclipse on 7 March 1970.** *Right:* **The observation room of the McMath-Pierce Solar Telescope.**

using a five-inch refractor with an enlarging lens which yielded an image of the Sun 20 inches in diameter, noted what he described as the *reseau photospherique*, or photospheric network, an appearance of smudged or forged areas among the granules. It was not clear whether this blurring was caused by currents in the Sun, in the Earth's atmosphere or perhaps even in the tube of Janssen's telescope.

The most important development in solar research was by means of the spectroscope, which was first applied to the study of the dark lines in the general solar spectrum, and the presence in the Sun of many elements additional to those found by Kirchhoff was demonstrated. However, at the 1868 eclipse of the Sun, spectra of the prominences—seen to be projecting parts of a continuous solar envelope named the 'Chromosphere'—were found to consist of bright lines that indicated a gaseous constitution. Soon after this eclipse, Janssen, and Sir **J Norman Lockyer (1836-1920)**, independently devised a method of seeing these bright lines in full daylight. Following on this, **Sir W Huggins** and **JKF Zollner (1834-1882)** showed that if the slit of the spectroscope is widened, the prominences themselves could be seen without any eclipse. An increasing number of bright prominence lines were noted at eclipses, with nine visually in 1868 and 29 photographically in 1878. Of these, a bright prominence near the D line of sodium was later identified as due to a gas—helium—discovered terrestrially by Ramsay in 1895.

The spectrum of the corona was seen in 1868 as a faintly continuous one, and in 1869 by **CA Young (1834-1908)** as continuous, with a bright green line from an unknown source which was given the name 'coronium,' although it was later found to be due to highly ionized iron. In 1893, the Sun's spectrum was noted to be distinct from that of the chromosphere or the prominences, and to have at least seven other bright lines of a wavelength shorter than that of the line in the green. Before the end of the nineteenth century, studies by AC Ranyard and others of the forms of the corona, as seen

THE VALUES OF THE SUN'S DISTANCE CURRENT AT VARIOUS TIMES

Date	Average Distance in Miles	Theoretician
140 AD	4,500,000	Ptolemy
1620	13,500,000	Kepler
1660	20,000,000	Hevelius
1700	86,000,000	Cassini
1700	81,700,000	Flamsteed
1820	95,200,000	Delambre
1835	95,400,000	Encke
1860	91,600,000	Hansen, Le Verrier
1900	92,900,000	International Agreement
Today	92,900,000	International Agreement

SOLAR ASTRONOMY 77

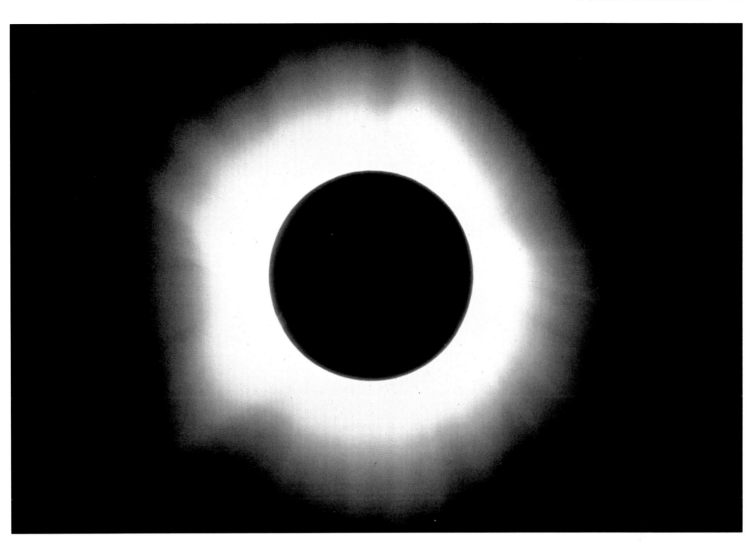

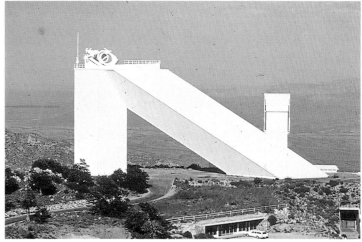

or photographed at different eclipses, showed that it varied in shape with the sunspot cycle, extending out in all directions at sunspot maximum, but equatorial extensions predominating at minimum. Prominences also seem to follow the same cycle of change.

During the solar eclipse of 22 December 1870, the outstanding event was the spectroscopic discovery by Young of the **Reversing Layer**. Young said that 'As the crescent grew narrower, I noticed a fading out, so to speak, of all the dark lines in the field of view, but was not at all prepared for the beautiful phenomenon which presented itself when the Moon finally covered the whole photosphere. Then the whole field was at once filled with brilliant lines, which suddenly flashed into brightness and then gradually faded away until, in less than two seconds, nothing remained but the

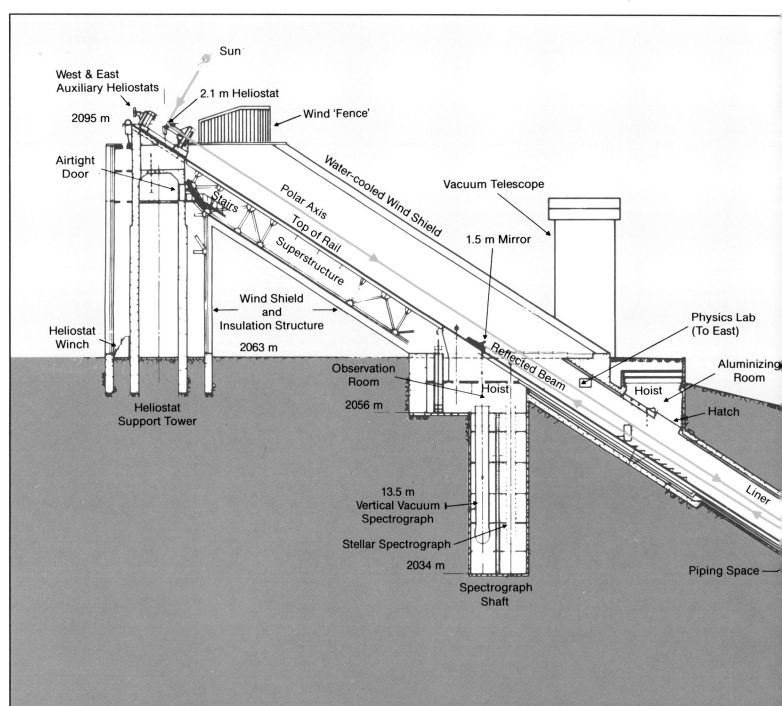

THE McMATH-PIERCE SOLAR TELESCOPE

lines I had been watching.' The impression produced was a complete reversal of the dark Fraunhofer lines into the bright lines of the gaseous absorbing stratum at the Sun's surface, which causes the absorption of dark lines of the solar spectrum.

The ideas of the French astronomer **HAE Faye (1814-1902)** on the general theory of the Sun's constitution may be taken as representing the most advanced opinion of his time. He believed that the Sun was mainly a gaseous body, that its radiation was due to the transport of heat upward by convection currents with cooler matter descending, that the photosphere was a surface of condensation at the outer limits and that sunspots were breaks in photospheric clouds. Only the first of these ideas survived later research.

NASA satellites demonstrated that Earth's tenuous upper atmosphere is not as stable and quiescent as previously believed. It swells by day and contracts at night. Its volume and density wax and wane with such solar events as solar flares, the 11-year solar cycle, and the 27-day solar-rotation period. NASA spacecraft confirmed the existence of the solar wind. They discovered that the solar wind streams outward along the Sun's magnetic field lines. Analyzing magnetic field and solar wind data from spacecraft near and far, scientists have redrawn the picture of interplanetary space. The solar wind and solar magnetic field are now visualized as forming a vast heliosphere encompassing space billions of kilometers outward from the Sun. The **Solar Mesosphere Explorer**, launched 6 October 1981, provided comprehensive data on how solar radiation creates and destroys ozone in the mesosphere, an atmospheric layer below the thermosphere and above the stratosphere. The University of Colorado designed and built the Solar Mesosphere Explorer and operated it for a year after launch.

Three **International Sun-Earth Explorers (ISEE)** were a joint project of NASA and the European Space Agency. ISEE 1 and 2 were launched into Earth orbit on 22 October 1977, and ISEE 3 had been placed in a heliocentric orbit near the Sun-Earth libration point on 12 August 1977. The ISEE program focused on solar-terrestrial relationships as a contribution to the International Magnetospheric Study. The three spacecraft obtained a treasure trove of new information on the dynamics of the geomagnetosphere, the transfer of energy from the solar wind, and energization of plasma in the geomagnetotail. For example, ISEE 1 found ions from our ionosphere accelerated in the geomagnetotail to fairly high energies. Previously, scientists thought these high energy particles originated from the solar wind.

In 1982, when its mission was completed with fuel to spare and all instruments in working condition, ISEE 3 was put through a series of complex maneuvers to explore the Earth's magnetotail through December 1983, to fly across and study

Far left: **Kitt Peak's McMath-Pierce Solar Telescope.** *Left:* **A diagram of the telescope, a portion of which is underground. The light shaft is inclined at 32 degrees to the horizontal and pointed at the celestial North Pole.** *Below:* **The heliostat mirrors.**

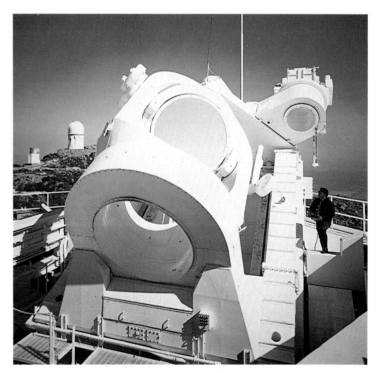

the wake of Comet Giacobini-Zinner in September 1985, and to observe from comparatively close range the effects of the solar wind on Halley's Comet in late 1985 and early in 1986. ISEE 3 was the first spacecraft to fly through a comet's tail and to survey Earth's long magnetotail. NASA's **Orbiting Solar Observatories (OSO)**, using X-ray and ultraviolet sensors, acquired a rich harvest of solar data during the 11-year solar cycle when solar activity went from low to high and then back to low. They photographed for the first time the birth of a solar flare, a great outburst of matter and energy from the Sun. When directed toward Earth, solar flares can cause blackouts of communications and electricity, crazily spinning magnetic compasses and enhancing displays of the northern and southern lights. OSOs discovered evidence of gamma radiation resulting from solar flares, indicating nuclear reaction in the flares.

OSO 1, launched on 7 March 1962, showed a correlation between fluctuations in temperatures of Earth's upper atmosphere and variations in solar ultraviolet ray emissions. **OSO 3**, launched on 8 March 1967, revealed that the center of our galaxy was the source of intense gamma radiation, which, when confirmed by the **High Energy Astronomy Observatory (HEAO) 3**, led to speculation that matter and antimatter were annihilating each other, leaving energy in the form of gamma rays. The data from **OSO 5**, launched 22 January 1969, revealed that Earth's upper atmosphere may contain as much as 10 times the amount of deuterium (a form of hydrogen with twice its mass) as previously estimated.

The Orbiting Solar Observatories discovered and provided information on solar poles where there were cooler and thinner gases than in the rest of the corona. Later, they discovered comparable phenomena on other areas of the Sun and scientists named them solar holes. They observed and reported on the dramatic coronal transient, a solar explosion hurling out hundreds of thousands of tons of material in huge loops at millions of kilometers per hour.

Infinitesimal (.001) reductions in solar energy output that may be related to unusually harsh winters and cool summers on Earth were discovered by the **Solar Maximum Mission (SMM)** satellite. 'Solar Max' was launched 14 February 1980, to study the Sun during the high part of the solar cycle. SMM also made the first clear observations of neutrons traveling from the Sun to the Earth after a flare and confirmed that fusion, the basic process that powers the Sun, occurs in the solar corona during a flare. A malfunction in January 1981 cut SMM's life short. Solar Max was recovered aboard the Space Shuttle orbiter *Challenger* during Mission 41-C in April 1984 and returned to Earth for repair.

Sun-orbiting Pioneers have contributed volumes of data about the solar wind, solar magnetic field, cosmic radiation, micrometeoroids, and other phenomena of interplanetary space. **Pioneer 4**, launched 11 March 1960, confirmed the existence of interplanetary magnetic fields and helped explain how solar flares trigger magnetic storms and the northern and southern lights (auroras) on Earth. The satellite also showed that the Forbush decrease of intergalactic cosmic rays near Earth after a solar flare was the same in interplanetary space, and thus does not depend upon an Earth-related phenomenon, such as the geomagnetic field.

Pioneers 6 through 9, launched 16 December 1965, 17 August 1966, 13 December 1967 and 8 November 1968, supplied volumes of data on the solar wind, magnetic and electrical fields, and cosmic rays in interplanetary space. Pioneer 7 detected effects of Earth's magnetic field more than three

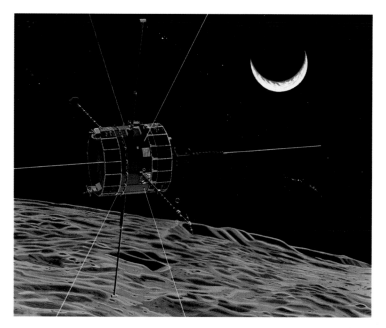

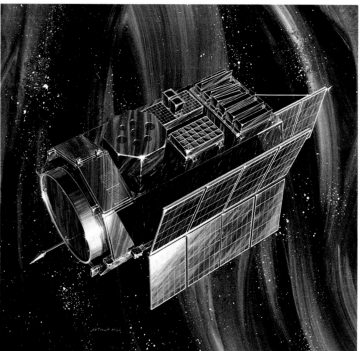

At top: **ISEE 3.** *Above:* **HEAO 3.** *Right:* **Two spectrometers aboard the OSO 8 conduct structure studies of the chromosphere and take high resolution ultraviolet spectrometer measurements.**

million miles outward from the night side of Earth. Pioneer 6 to 9 data also drew a new picture of the Sun as the dominant phenomenon of interplanetary space. They found that the solar wind continues well beyond the orbit of Mars. (**Pioneer 10**, the first Jupiter explorer, continued to report its existence as it crossed the orbit of Pluto.)

Analyses of their data indicated that the solar wind is an extension of the solar corona, the Sun's atmosphere. They revealed that the wind draws out the Sun's magnetic field to form what were previously called interplanetary magnetic fields but now is referred to as the heliosphere. They showed that the combination of the solar wind's outward pull and the rotation of the Sun caused the lines of forces of the magnetic field to be twisted like streams of water from a whirling lawn sprinkler. Pioneer data showed that solar cosmic rays spiral around the lines of force of the Sun's magnetic field. This indicates they travel through space in well-defined streams.

NASA Space Science

The American **National Aeronautics & Space Administration (NASA)** was formed in October 1958 as an umbrella organization for all American efforts to explore space. NASA's space science programs have contributed significantly to a new golden age of discovery. They have substantially advanced the frontiers of knowledge about our home planet, the relationships of Sun and Earth and celestial phenomena. NASA satellites discovered the existence of the **Van Allen Radiation Region** around Earth. They demonstrated that Earth's magnetic field is not shaped like iron filings around a bar magnet but like a vast cosmic teardrop. It is literally blown into this shape by the solar wind, that hot electrified gas constantly speeding out from the Sun. Satellites also contributed to the understanding of interaction of solar activity with the magnetic field to produce periodic radio blackouts, trip circuit breakers of electric transformers, cause magnetic compasses to become erratic, and generate auroras.

Satellites have dramatically altered our conception of the universe. Our atmosphere blocks most of the electromagnetic radiations that can tell us about the nature of celestial objects. Satellite observatories viewing the heavens from above our appreciable atmosphere open a window on the universe.

Astronomers had theorized that only a small fraction of the electromagnetic radiation emitted by stars consisted of X-rays. (X-rays are generated by high energy processes; a sparse emission of X-rays would indicate that our universe was comparatively peaceful and slowly evolving.) However, NASA satellites revealed a skyful of X-ray sources. This revolutionized the concept of the universe to one whose dynamics and evolution are governed by dramatic and enormously powerful processes. Our study of the universe draws us toward the answers to fundamental questions about the very nature of matter, life and the destiny of the stars.

Arguably, the most dramatic discoveries have resulted from observations of other planets and, if present, their moons. The observations accentuated the uniqueness of our planet, its resources and its environment. It is a place where a variety of suitable conditions combined to create and sustain life.

Our nearest neighbor, the Moon, is a radically different world. NASA's Apollo manned expeditions and unmanned

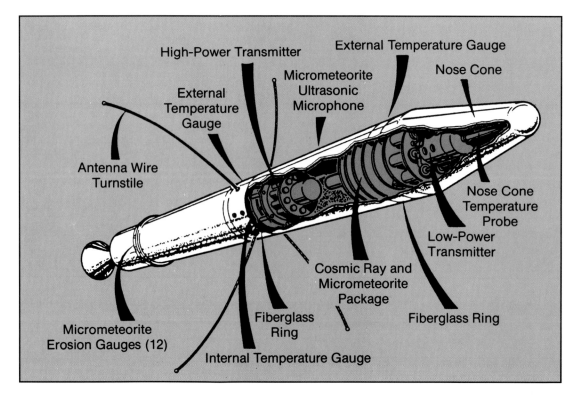

These pages: **Explorer 1, the first of many small unmanned satellites to conduct space science missions, satellite applications and technological research.**

spacecraft, as well as telescope observations, show it to be pockmarked with huge meteorite craters partially filled with basalts from ancient lava flows. Its surface is wracked with excessive heat by day and cold by night. It is bombarded by solar radiation because it has neither an intrinsic atmosphere nor magnetic field to ward off the radiation. It has no signs of life nor even evidence that life processes have begun.

Since 1 February 1958, when two organizations that later became part of NASA launched **Explorer 1**, NASA has launched more than 60 Explorers. A series of satellites that was comparatively small and varied in size and shape, it carried a limited number of experiments into the most suitable orbits. The Explorer designation has been assigned not only to small satellites conducting space science missions but also to those whose primary mission is in either satellite applications or technological research. Explorers devoted principally to space science programs are described in this section.

NASA's own first successful satellite launch was **Explorer 6**, orbited 7 August 1959. Explorer 6 added to information about the Van Allen Region and micrometeoroids. It also telecast a crude image of the north Pacific Ocean. **Explorer 7**, launched 13 October 1959, provided data revealing that the Van Allen Region fluctuated in volume intensity and suggested a relationship of the region with solar activity. It indicated that variations in solar activity may also be related to the abundance of cosmic radiation, magnetic storms and ionospheric disturbances in Earth's vicinity.

Explorer 10, launched on 25 March 1961 to gather magnetic field data, was the first spacecraft to obtain information that suggested that the interplanetary magnetic field may actually be an extension of the Sun's field carried outward by the solar wind.

NASA's launch of **Explorer 11** on 27 April 1961 was the first step of its long range program to probe the universe's secrets that are veiled by Earth's atmosphere. Designed to monitor gamma rays, the satellite's data appeared to contradict the steady state theory of constant destruction and creation of matter.

With **Explorer 12**, launched 15 August 1961, scientists were able to arrive at many conclusions about space: the Van Allen Region is a single system of charged particles rather than several belts, Earth's magnetic field has a distinct boundary, the solar wind compresses the Earth's magnetic field on the Sun's side and blows it out on the other, and geomagnetic storms that cause radio blackouts and power outages may result from solar flares.

On 2 October 1962, NASA launched **Explorer 14** to monitor the Van Allen Radiation Region during a period of declining solar activity. Scientists in the meantime discovered that the United States project Starfish, involving a high altitude nuclear burst in July 1962 had created another artificial radiation belt. Explorer 14 was joined by **Explorer 15** on 27 October to help monitor this belt. The two satellites' data helped ease scientific anxiety by indicating that atomic particles making up the belt were rapidly decaying.

NASA launched **Explorer 26** on 21 December 1964, and its data increased understanding of how atomic particles traveling toward Earth from outer space are trapped by Earth's magnetic field and how they spiral inward, along Earth's magnetic field lines in the northern and southern latitudes, interacting with the atmosphere to generate auroras.

NASA and the US Navy launched **Explorer 30** and **Explorer 37** on 19 November 1965, and 5 March 1968, respec-

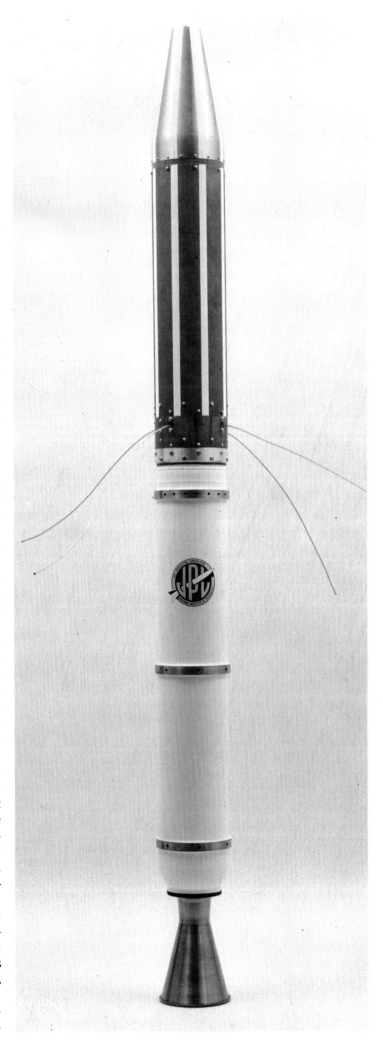

tively. The satellites monitored solar X- and ultraviolet rays during periods of declining and increasing solar activity. Perhaps the longest satellites ever launched were NASA's Radio Astronomy **Explorer 38** and **Explorer 49**. The tiny bodies of these satellites were each crossed with two antennas that were about three times as long as the Washington Monument is high. The antennas were unreeled to form a vast 'X' in space where they received natural radio signals that do not ordinarily reach Earth, thus filling a gap in our radio astronomy knowledge.

Explorer 38, launched on 4 July 1968, surprised astronomers by reporting that Earth sporadically emits natural radio waves. Until then, the only planet known to emit radio waves was Jupiter. Earth was so noisy that it drowned out many other sources. However, **Explorer 35** was also able to report that the Sun emitted more low frequency radio signals than scientists anticipated. Earth's radio noise prompted NASA to make Explorer 49, the second **Radio Astronomy Explorer**, into a lunar-anchored satellite. After the 10 June 1973 launch, Explorer 49 was maneuvered into lunar orbit, far enough away to prevent radio interference from Earth.

Explorer 42, launched 12 December 1970, was the first of a new category of NASA spacecraft called **Small Astronomy Satellites (SAS)**. Designed to pick up X-rays, it gathered more data in a day than sounding rockets accumulated in the nine previous years of X-ray astronomy. Astronomers used its data to prepare a comprehensive X-ray sky map and X-ray catalogue. Data from Explorer 42 suggested that superclusters of galaxies may be bound together by tenuous gases whose total mass is greater than that of the optically visible galaxies. This would provide a significant percentage of the mass needed to support the theory that our expanding universe will eventually contract. Explorer 42 was the first NASA satellite launched by a foreign nation. Italy launched Explorer 42 from its floating San Marco platform in the Indian Ocean off the coast of Africa, near Kenya. Because Explorer 42 was launched on Kenya's independence day, it was also named 'Uhuru' which is Swahili for 'freedom.'

On 15 November 1972, Italy launched NASA's **Explorer 48**, the second SAS, from its San Marco platform. Explorer 48 continued the expansion of knowledge about gamma ray sources first begun by Explorer 11. It provided data that could be interpreted as supporting the theory that the universe is composed of regions of matter and antimatter. **Explorer 53**, the third SAS, was launched from San Marco on 7 May 1975. It discovered many additional X-ray sources, including one identified as a quasi-stellar object (quasar) only 783 million light years away. This is the closest quasar yet discovered.

Explorer 45, launched 15 November 1971, investigated the relationships of geomagnetic storms, particles radiation and auroras. It was the second satellite launched by an Italian crew from the San Marco platform off Kenya in the Indian Ocean.

NASA's **Interplanetary Monitoring Platforms**, or **IMP Explorers**, added significantly to knowledge about how Earth's magnetic field and the Van Allen Radiation Region fluctuate during the 11-year cycle. **IMP Explorer 18** confirmed that Earth's magnetic field was shaped like a giant cosmic tear drop. It discovered a shockwave ahead of the Earth's field. The shockwave is caused by the impact of the speeding solar wind with the Earth's field. Between the shockwave and the magnetopause, or magnetic field boundary, Explorer 18 discovered a turbulent region of magnetic fields and atomic particles.

Left: **Explorer 38 monitored low frequency radio signals in space.** *Above:* **Explorer 48 is ready for vibration testing.**

IMP Explorer 33 was the first satellite to provide evidence that the geomagnetic field on Earth's night side extends beyond the Moon. **IMP Explorer 35**, a lunar-anchored (lunar orbiting) IMP, gathered data about micrometeoroids, magnetic fields, the solar wind, and radiation at lunar altitudes. Its instruments revealed the Moon to be what one scientist termed a 'cold nonmagnetic nonconducting sphere.'

Many new discoveries were made by the **International Ultraviolet Explorer (IUE)**, launched 26 January 1978. IUE was a joint project of NASA, the United Kingdom and the European Space Agency. IUE data supported a theory that a black hole with the mass of a thousand solar systems existed at the center of our Milky Way galaxy and revealed that our galaxy had a halo of hot gases. The data provided evidence that so-called twin quasars were actually a double image of the same object. Light waves from the quasar are bent around a massive elliptical galaxy which acts as a gravitational lens to produce the double image picked up by ground observatories.

Atmosphere Explorers confirmed or redrew our conceptions of Earth's tenuous upper atmosphere. The first, **Explorer 8**, was launched 3 November 1960. It confirmed that temperatures of electrons in the upper ionosphere are higher by day than by night. It discovered that oxygen predominates in the ionosphere up to an altitude of about 650 miles where helium predominates. A secondary experiment indicated that micrometeoroid quantities varied inversely with size.

Air Density Explorer 9, **Air Density Explorer 24** and **Air Density Explorer 39** were launched 16 February 1961, 21 November 1964 and 8 August 1968, respectively. These were essentially 12-foot balloons of aluminum foil and plastic laminate that were inflated in orbit. Air drag on the satellite indicated air density. The satellites revealed that atmospheric density varied from day to night, with the 27-day rotation period of the Sun, with the 11-year solar cycle, and with violent eruptions on the Sun. **Explorer 25** was launched on the same Scout booster that orbited **Explorer 24**, the first multiple launch by a single vehicle. **Explorer 40** was launched with the same Scout vehicle as Explorer 39. They demonstrated a correlation between air density and solar radiation. Explorers 25 and 40 were called Injuns and University Explorers because they were built by the University of Iowa.

Explorer 17, launched 2 April 1963, and **Explorer 32**, launched 25 May 1965, gathered information about the composition of neutral atoms and molecules. Explorer 17 confirmed Explorer 8 indications of a belt of neutral helium in the upper atmosphere. In radio-echo soundings of the ionosphere, radio signals at different frequencies were transmitted from the ground. The reflected frequency discloses electron density, and the return time indicated the altitude or distance at which the density was encountered. Ground-based sounding cannot provide information about the upper ionosphere because electron density increases up to a certain altitude and then tapers off. In addition, many areas of Earth are too remote or inaccessible for ground-based radio sounding. These problems were solved by using satellites as topside sounders. They beamed radio waves into the ionosphere from altitudes far above the region of maximum electron density.

NASA topside sounders included **Explorer 20**, launched 25 August 1964, and **Explorer 31**, launched 28 November 1965. Scientists correlated data from these satellites with data

Left: Explorer 17 photographed in a space environment. *Above:* Explorer 20 (shown with protective black coating) after the nose fairing is jettisoned with a 4-inch experiment sphere atop the truncated cone. *Right:* The Interplanetary Monitor Platform satellite Explorer 18 (IMP-1).

from the Canadian topside sounders **Alouette 1** and **2**. **Explorer 22**, an ionosphere beacon launched 10 October 1964, also measured electron density in the ionosphere. Explorer 22 was built with quartz reflectors for the first major experiment in laser tracking.

The thermosphere, a region of the upper atmosphere, was believed to be relatively stable until **Explorer 51**, **Explorer 54** and **Explorer 55** were launched. These satellites were equipped with onboard propulsion systems that enabled them to dip deep into the atmosphere and pull out again, taking measurements and providing extensive data about the upper thermosphere. They found that the thermosphere behaved unpredictably with winds 10 times stronger than normally found at Earth's surface. They discovered abrupt and constantly changing wind shears. Their data contributed significantly to knowledge about energy transfer mechanisms and photochemical processes (such as those that create the ozone layer) in the atmosphere. Launch dates for the three Explorers were 16 December 1973, 6 October 1975 and 19 November 1975.

A two-satellite **Dynamic Explorer** project, launched simultaneously on 3 August 1981, significantly contributed to data on coupling of energy, electric currents, electric fields, and plasmas (hot electrified gases) between Earth's magnetic field, the ionosphere, and the rest of the atmosphere. Among their discoveries were nitrogen ions in the geomagnetosphere. They also confirmed the existence of the polar wind which is an upward flow of ions from the polar ionosphere.

Man's view of the universe is narrowly circumscribed by the atmosphere which blocks or distorts most kinds of electromagnetic radiations from space. Analyses of these radiations (radio, infrared, visible light, ultraviolet, X-rays and gamma rays) give important new information about the phenomena in our universe. The small Astronomical Explorers indicated the great potential for acquiring new knowledge by placing instruments above Earth's obscuring atmosphere. Consequently, NASA orbited a series of large astronomical observatories bearing a great variety of instruments which have significantly widened our window on the universe.

The first successful, large-scale observatory was **Orbiting Astronomical Observatory 2 (OAO 2)**, nicknamed *Stargazer*, which was launched 7 December 1968. In its first 30 days, OAO 2 collected more than 20 times the celestial ultraviolet data acquired in the previous 15 years of sounding rocket launches.

Launched 21 August 1972, **OAO 3** was named for the Polish astronomer Copernicus. The satellite provided much new data as it continued studies of the outer atmospheres of Earth, Mars, Jupiter and Saturn. It gathered data on the black hole candidate Cygnus X-1, so named because it was discovered in the constellation Cygnus.

Much of its data supported the hypothesis that Cygnus X-1 is a black hole. A black hole is a one-time massive star that has collapsed to such density that it does not permit even light or other electromagnetic radiations to escape it. Scientists can study Cygnus X-1 because it is part of a binary star system and has a visible companion. In addition, according to theory, a substantial part of the matter dragged into a black hole is transformed into X-rays and gamma rays that are radiated into space before they reach the point of no return.

OAO 2 DISCOVERIES

■ Stars that are many times as massive as our Sun are hotter and consume their hydrogen fuel faster than estimated on the basis of ground observations. OAO 2 data contributed to resolving a disparity between observations made from the ground and theories of stellar evolution.

■ Another stellar theory was brought into question. According to this theory, intensity of celestial objects should be less in ultraviolet light than in visible light. However, several galaxies that looked dim in visible observations from Earth were bright in ultraviolet observations by *Stargazer*.

■ Diffuse dust nebulae are regarded by many astronomers as the location for the formation of stars.

■ *Stargazer* was able to observe Nova Serpentis in 1970 for 60 days after its outburst. It confirmed that mass loss by the nova was consistent with theory.

■ *Stargazer* observations of the Comet Tago-Sato-Kosaka supported the theory that hydrogen is a major constituent of comets. It detected a hydrogen cloud as large as the Sun around the comet. Because of our atmosphere, this hydrogen cloud could not be detected by ground observatories.

■ Looking toward Earth, *Stargazer* reported that the hydrogen in Earth's outer atmosphere is thicker and covers a larger volume than previous measurements indicated.

Left above: **The Dynamics Explorer-A (DE-A) undergoing preflight evaluation testing. Together with a sister spacecraft, the two will explore a boundary region between Earth and space that affects the atmosphere, auroral displays, radio transmissions, climate and weather.** *Left below:* **Atmosphere Explorer 51 provided data on the thermosphere that greatly surpassed prior knowledge obtained by sounding rockets.** *Right:* **Explorer 47, the ninth IMP was launched 22 September 1972 aboard a three-stage Delta rocket. The automated space physics laboratory conducted studies of interplanetary radiation from an orbit approximately half way to the Moon.**

OBSERVATIONS MADE BY OAO 3 (COPERNICUS)

- Interstellar dust clouds have fewer heavy elements than our Sun. This supported the contention that the Sun and planets coalesced from the debris of an ancient supernova.
- Larger amounts of hydrogen molecules than expected were found in interstellar dust clouds.
- Surprisingly large amounts of deuterium (an element with the same atomic number and the same position in the Table of Elements as hydrogen but with twice the mass) were also detected in interstellar dust clouds.

The latter two findings suggest that star formation may be common. The discovery regarding deuterium contradicted a theory that most deuterium, a basic element for atomic fusion in stars, has been exhausted.

Launched on 25 January 1983, the **Infrared Astronomy Satellite (IRAS)** revealed many infrared sources in the Large Magellanic Cloud, 155,000 light years from Earth, that are not visible from Earth, helping scientists compile the first catalogue of infrared sky sources. Because all objects, even cool dark ones that may be the black cinders of dead stars, radiate infrared light, IRAS may discover many other invisible objects. It has revealed stars being born in thick opaque clouds of gas. IRAS is a joint project of NASA, the Netherlands and the United Kingdom. In April 1953, IRAS detected a new comet which came within 3 million miles of Earth in May, the closest comet approach in 200 years. It most recently discovered a possible new solar system near the star Vega.

In June 1991, astronomers studying very faint objects originally detected by the IRAS discovered a new, very distant object which was considered to be the most luminous object ever seen in the universe. These observations, published on 27 June in the British science journal *Nature*, showed that this object was a massive dust cloud which radiated 99 percent of its light in the infrared part of the spectrum. The team of astronomers believed that this mysterious cloud may have been a massive galaxy in the process of formation, or alternatively, a quasar embedded in the dust of a massive galaxy.

Infrared light—or, more simply, heat radiation—is invisible to the human eye but can be detected by electronic sensors, such as those on the IRAS satellite. Often these sources turn out to be dusty objects, because dust particles are very efficient emitters of infrared radiation. IRAS discovered hundreds of thousands of infrared objects that astronomers continued to observe more closely in order to determine what they were.

In June 1992, **Dr Kenneth Marsh** and **Dr Michael J Mahoney**, two astronomers at NASA's Jet Propulsion Laboratory at Pasadena, while studying IRAS data, discovered evidence of planets, or other bodies, around eight stars in a star-forming region of the Milky Way Galaxy, 450 light years from Earth. The scientists said their discovery of unseen companions around low-mass stars in the Taurus-Auriga region of the Milky Way resulted from the study of data from IRAS acquired in 1983 and data from ground-based observatories acquired between 1981 and 1983.

Taurus-Auriga is a giant gas and dust cloud complex, further out from the center of the galaxy than Earth's Solar

Above: **IRAS.** *Below:* **An infrared IRAS photograph of the Milky Way and our galactic center. Our Solar System takes 225 million years to revolve around this center.** *Right:* **A proto star in Barnard 5.** *Right below:* **The Orion Nebula seen by IRAS.**

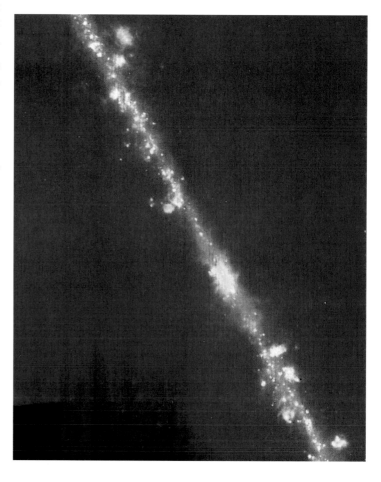

System, and is one of two star-forming regions closest to Earth. The stars studied were called T Tauri stars, which have about the same mass as the Sun. T Tauri stars are believed to be young and are usually found in groups embedded in clouds of gas and dust. Earth's Solar System is believed to have formed similarly, with the planets accreting from the matter in the solar disc left over after the formation of the Sun. By looking in the infrared part of the spectrum at the variations of emissions from the T Tauri solar disc, the two scientists found evidence of gaps, indicating that a body had accreted and swept an orbit around a young star, forming a gap in the disc.

They identified eight such stars. In some cases, wide gaps were found, suggesting a fairly massive companion, such as a low-mass star or brown dwarf. A brown dwarf is a body that might have become a sun if it had sufficient mass for nuclear reactions to begin in the core. For three stars, however, the gaps were much narrower, indicating many fewer massive companions which could be planets. These three stars were T Tauri, HK Tauri and UY Aurigae.

Looking at gaps in the stellar discs is a way of finding evidence of star companions, the scientists said. Previously, astronomers looked at the pertubations of some stars to see if the 'wobble' of the stars could possibly be caused by the gravitational influence of nearby bodies. Marsh said that although the data they used also was studied by other scientists, they were more interested in star formation and were not looking for planets. 'But we are in an environment where people are very interested in planets,' he explained.

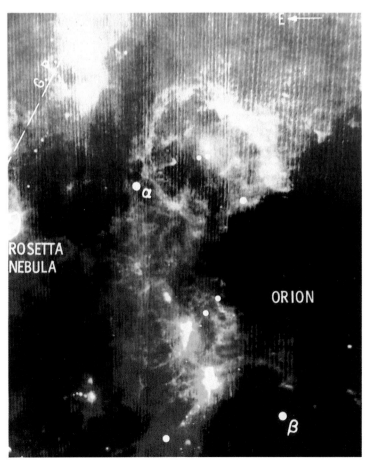

Above: **HEAO 3 scaned the sky for cosmic and gamma ray sources.** *Below:* **OGO-B in NASA's Space Environment Simulator.**

NASA's three **High Energy Astronomy Observatories (HEAOs)** portray a universe in constant turbulence with components repeatedly torn apart and recombined by violent events. **HEAO 1**, launched 12 August 1977, also discovered a new black hole candidate near the constellation Scorpius, bringing the total to four. Other black hole candidates are in or near the constellations Cygnus, Circinus and Hercules.

Another major result of HEAO 1 was the superhot superbubble of gas 1200 light years in diameter and about 5000 light years from Earth. Centered in the constellation Cygnus, the bubble has enough gas to create 10,000 suns. HEAO 1 raised the catalogue of X-ray sources from 350 to about 1500.

The *Einstein* Observatory, nicknamed for the famous mathematician, is **HEAO 2**, launched 13 November 1978. *Einstein* was equipped with more sensitive instruments than HEAO 1. Thus, it was able to discern that the X-ray background observed by HEAO 1 was not coming from diffuse hot plasmas but from quasars. Einstein also provided the first pictures of an X-ray burster which is apparently located at the center of a globular cluster called Terzan 2. The bursters are frequently associated with clusters of old stars. They are usually explained in terms of gases interacting violently with neutron stars or black holes, emitting very short bursts of X-rays. Among other data from Einstein are X-ray spectra of supernova remnants. The data support the theory that our system was formed from debris of an ancient supernova.

HEAOs 1 and 2 X-ray measurements related principally to atomic interactions and plasma processes associated with stellar phenomena. **HEAO 3**, launched 20 September 1979, scanned the universe for cosmic ray particles and gamma radiation. The events that HEAO 3 measures result from nuclear reactions in the hearts of stellar objects and the elements they create. HEAO 3 has observed in the Milky Way's central region gamma rays that apparently emanate from the annihilation of electrons and positrons (the antimatter equivalent of electrons). It has also discovered an object emitting energy in the form of gamma rays equivalent to 50,000 times the Sun's total output. The object is 15,000 light years from Earth and appears to be undergoing processes on a comparatively small scale that are believed to occur in quasars on a large scale.

The **Netherlands Astronomical Satellite (NAS)**, launched 30 August 1974, was a small X-ray and ultraviolet orbiting observatory. Among discoveries from its data are X-ray bursters, sources that emit bursts of X-rays for seconds at a time. NAS was a cooperative program of NASA and the Netherlands.

A half dozen **Orbiting Geophysical Observatories (OGO)** launched between September 1961 and June 1965 have provided more than a million hours of scientific data from about 130 different experiments relating to Earth's space environment and Sun-Earth interrelationships. They significantly contributed to understanding of the chemistry of Earth's atmosphere, of Earth's magnetic field and of how solar particles penetrate and become trapped in the magnetosphere.

They provided the first evidence of a region of low energy electrons enveloping the high energy Van Allen Radiation Region, first observation of daylight auroras, first global map of airflow distribution, and much other knowledge about magnetic fields, particle radiation, Earth's ionosphere, the shockwave between the geomagnetic field and solar wind

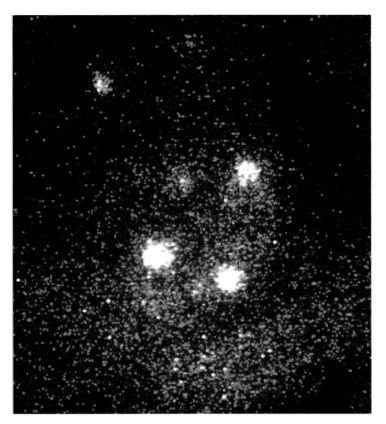
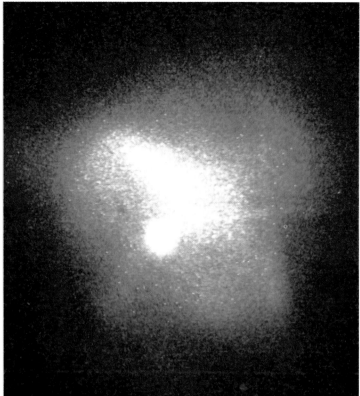

In these HEAO 2 photos, the bright spots correspond to regions of intense X-ray emission: Young stars (*above*), a remnant of an exploding supernova in Cassiopeia (*below*) and a pulsar in the Crab Nebula (*above*).

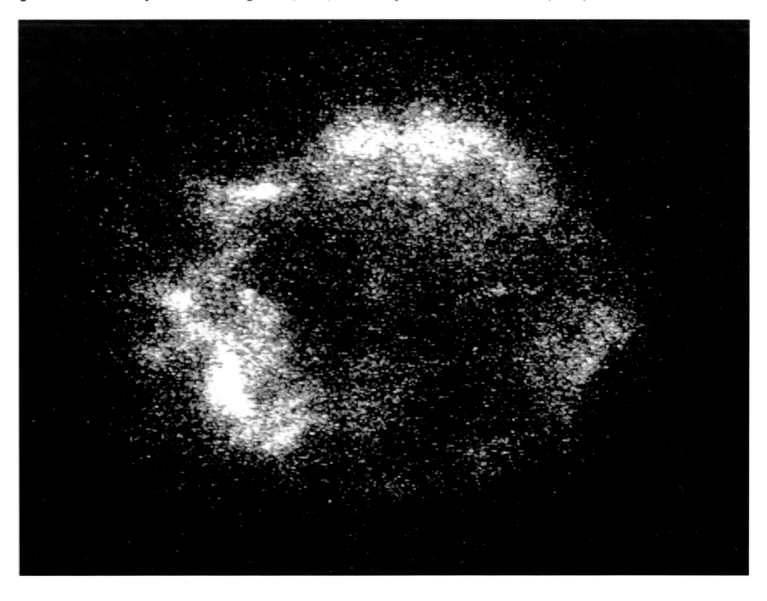

that was discovered by Explorer 18, and the hydrogen cloud enveloping Earth.

Beyond Earth, they completed the first sky survey of hydrogen, discovered neutral hydrogen around the Sun, found several strong sources of hydrogen in the Milky Way, and in April 1970, detected a cloud of hydrogen 10 times the size of the Sun around Comet Bennett. The existence of large amounts of hydrogen around comets was first discovered around comet Tago-Sato-Kosaka in January 1970 by Orbiting Astronomical Observatory 2 *Stargazer*. The large amounts of hydrogen around comets that were observed by OAO and OGO clearly establish that hydrogen is a major constituent of comets. **Mariner 2** was NASA's first successful interplanetary spacecraft. Launched 27 August 1962, it flew past, and made close-up observations of, Venus on 14 December 1962. Its data supported Earth-based microwave scans that suggested a surface temperature as high as 800 degrees Fahrenheit, hot enough to melt lead on both day and night sides. It detected no openings in the dense clouds enveloping Venus. Its data indicated no intrinsic Venusian magnetic field nor increase in radiation. This suggested that Venus has no radiation belt like the Van Allen Radiation Region around Earth. The data are consistent because the Van Allen Radiation Region is attributed to the capture of energetic particles by Earth's magnetic field. Mariner 2 also confirmed the predominance of the solar wind as a feature of interplanetary space and the ubiquity of interplanetary magnetic fields, which scientists now realize are an extension of the Sun's magnetic field dragged out into space by the solar wind.

Passing Mars on 14 July 1965, **Mariner 4**, launched 28 November 1964, gave the world its first close look at that planet's surface. The pictures were surprising and showed a heavily cratered moon-like surface that looked as though it may not have changed much in billions of years. Because the pictures covered about one percent of Mars, they permitted no conclusions until other spacecraft viewed additional areas of the planet. The pictures covered some areas crossed by the supposed Martian canals but showed no readily apparent straight-line features that could be interpreted as artificial. Among other data from Mariner 4 were indications that the surface atmospheric pressure on Mars was less than 10 millibars. Earth's sea level air pressure is about 1000 millibars. Humans would need a pressure suit on Mars. Mariner 4 also gave additional information about the planet's size, gravity and path around the Sun. Mariner 4 detected neither a Martian magnetic field nor radiation belt but revealed a Martian ionosphere.

Mariner 5 was launched 14 June 1967, to refine and supplement data about Venus obtained from Mariner 2 and other observations. It contained improved instrumentation and in October 1967, flew within 2500 miles of Venus as compared to the 21,645-mile closest approach of Mariner 2.

Mariners 6 and **7** were launched on 25 February and 27 March 1969, respectively, and flew as close as 2,000 miles to Mars on 31 July and 5 August 1969, respectively. Mariner 6 flew past the planet along its equator. Mariner 7 overlapped part of the Mariner 6 ground track and then sped south over the south polar ice cap. They took pictures of Mars and studied it with infrared and ultraviolet sensors. Their pictures show not only cratered but also smooth and chaotic surfaces.

Launched 30 May 1971, **Mariner 9** went into Martian orbit on 13 November 1971, the first spacecraft placed into orbit around another planet. It orbited and studied Mars and the planet's two tiny satellites, Deimos and Phobos, until 27

Left top: Mariner 2 spacecraft. *Left:* Mariner 10 spacecraft. *Above:* A view of Mariner 8, the twin of Mariner 9, shows the scan platform at bottom carrying two television cameras and experiments. The green dish is a high-gain antenna. The louvers are for temperature control, as is the white shroud covering the 300-pound thrust retro-engine. *Right:* Mariner 5.

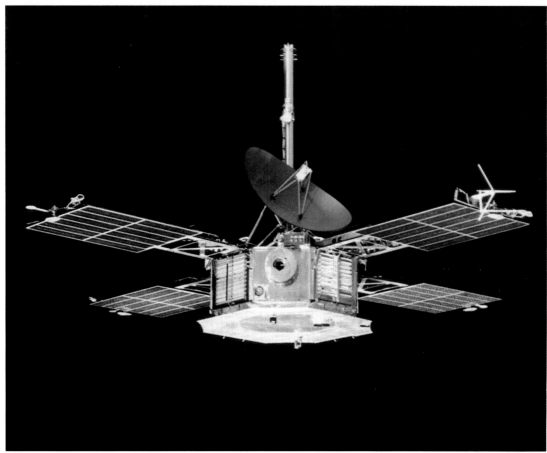

MARINER 5 VENUS OBSERVATIONS

- The atmosphere is at least 80 percent carbon dioxide.
- The atmosphere is about 100 times denser than Earth's.
- The surface temperature may be as high as 800 degrees Fahrenheit.
- The solar wind is diverted around Venus by the planet's ionosphere. (Earth's magnetic field diverts the solar wind around our planet.)
- The Venusian exosphere, like Earth's, is made up largely of hydrogen.
- Venus has no detectable magnetic field nor radiation belt.

October 1972. Mariner 9 arrived at a discouraging time when a dust storm enveloped most of the planet. Even the dust storm, however, provided information of value such as atmospheric circulation pattern and the fact that only on Mars were dust storms of such magnitude observed. When the storm cleared, Mariner 9 was able to photograph Martian geography in remarkable detail. Its photographs show Martian volcanic mountains, such as the 15.5-mile-high Olympus Mons, which is larger than any mountain on Earth. Also discovered was the vast, 3000-mile Valles Marinaris (Mariner Valley), long enough to stretch across the United States from the Atlantic to the Pacific Ocean; and signs that rivers, and possibly seas, may have existed on Mars.

Mariner 10, launched on 3 November 1973, flew by Venus on 5 February 1974 and in a solar orbit, swept nearby and gathered information about Mercury on three separate occasions: 29 March and 21 September 1974 and 16 March 1975.

Left: This view of Venus was taken by Mariner 10 on 6 February 1974, one day after the spacecraft flew past Venus enroute to Mercury. Processing the film with a blue filter enhances the ultraviolet markings on the planet's clouds. The predominant swirl is at the South Pole. *Right:* Mercury's northern limb as seen by Mariner 10 on its 29 March 1974 flyby. *Right below:* Mariner 9 transmitted this image of Mars' outstanding feature, Olympus Mons, or Mount Olympus, the dormant shield-type volcano standing 15.5 miles above the Martian surface.

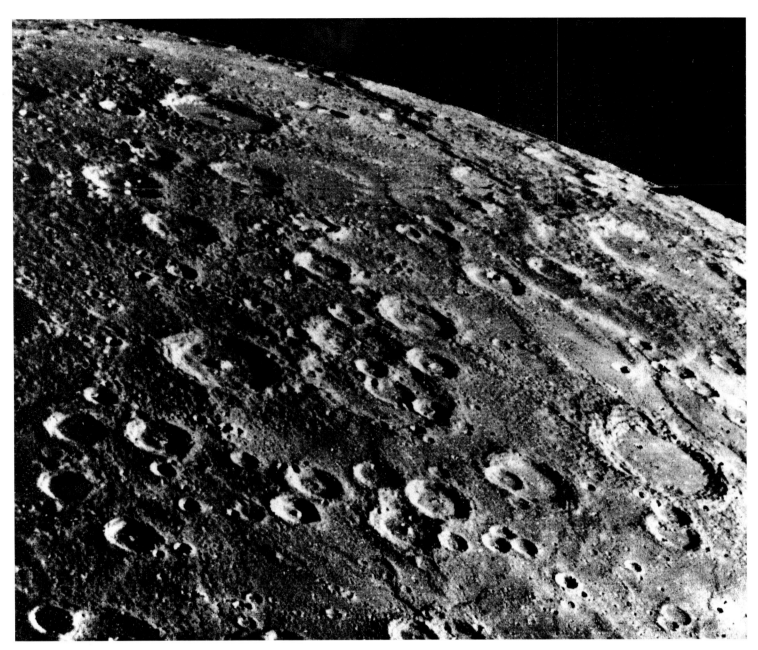

Johann Hieronymus Schroeter (1745-1816) became the first astronomer to record his observations of Mercury's surface detail, but his drawings, like those of **Giovanni Schiaparelli (1835-1910)** more than a century later, were ill-defined and turned out to be inaccurate. The great American astronomer **Percival Lowell (1855-1916)** reported that he had observed streaks on Mercury's surface similar to those that he and Schiaparelli had both observed on Mars. Schiaparelli had called these Martian features canali (channels) and Lowell decided they were *canals*, built by intelligent life. Both astronomers agreed, however, that the streaks on Mercury's surface were of natural origin. The major milestone in the observation of Mercury came in March 1974 when the American spacecraft Mariner 10 began a series of three flybys at a distance of about 12,000 miles, in which it was able to photograph, in great detail, objects as small as 325 feet across.

These first close-ups of Mercury reveal an ancient surface bearing the scars of huge meteorites that crashed into it billions of years ago. They show unique large scarps (cliffs) that appear to have been caused by crustal compression when the planet's interior cooled. A Mercurian magnetic field, about a hundredth the magnitude of Earth's, and an atmosphere, about a trillionth the density of Earth's, were detected. The Mercurian atmosphere is made up of argon, neon

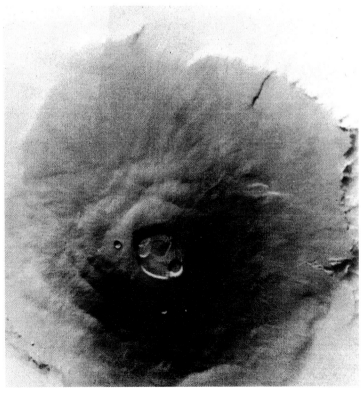

98 THE ILLUSTRATED GUIDE TO ASTRONOMY

and helium. Data suggest a heavy iron-rich core making up about half the planet's volume. (Earth's core is about 25 percent of its volume.) Mariner 10 reported that Mercury's surface temperatures were 950 degree Fahrenheit on the sunlit side and -350 degree Fahrenheit on the night side.

Mariner 10 was also the first imaging spacecraft to view Venus. The spacecraft's ultraviolet cameras revealed that Venus' topmost clouds circled the planet 60 times faster than Venus rotates. Mariner 10 confirmed a long-held theory about Earth's weather, namely that solar heat causes air to rise in the tropical area, flow to the poles, cool, fall and then return to the tropics where the process is repeated. No such process could be discerned on Earth, as Earth's rapid rotation, variable atmospheric water content, sizable axial tilt and mixing of continental and ocean air masses produce strong air currents obscuring the equatorial-polar flow. Venus has practically no water vapor, rotates slowly, has no axial tilt, and no mixing of continental and ocean air masses (no oceans) to obscure this flow.

Pioneer 10, launched 3 March 1972, was the first spacecraft to cross the Asteroid Belt, first to make close range observations of the Jupiter system, and the first to go beyond the outermost planets. In December 1973, it swept nearby Jupiter, finding no solid surface under the thick and deep clouds enveloping the planet. Thus, the world learned that Jupiter is a planet of liquid hydrogen. It explored the huge Jupiter magnetosphere, made close-up pictures of the Great Red Spot and other atmospheric features, and observed and measured at relatively close range Jupiter's large Galilean satellites—Io, Europa, Ganymede and Callisto.

Passing Jupiter, Pioneer 10 continued to map the heliosphere, the giant solar magnetic field drawn out from the Sun by the solar wind. Pioneer 10 found that the heliosphere, like the magnetospheres of Earth and Jupiter, behaves like a cosmic jellyfish, altering its shape in response to rises and falls in solar activity. It also reported that the speed of the solar wind does not decrease with distance from the Sun. On 13 June 1983, Pioneer 10 crossed the orbit of Neptune which will be the planet farthest from the Sun until the turn of the century. (This is because Pluto, although farthest on average, has an extremely elliptical orbit that crosses and goes inside of Neptune's.)

Below: **Pioneer 10 was the first outer Solar System probe and left the Solar System in 1985. Pioneer 11** *(at right)* **used 'gravity assist' in its encounters with Jupiter and Saturn.**

100 THE ILLUSTRATED GUIDE TO ASTRONOMY

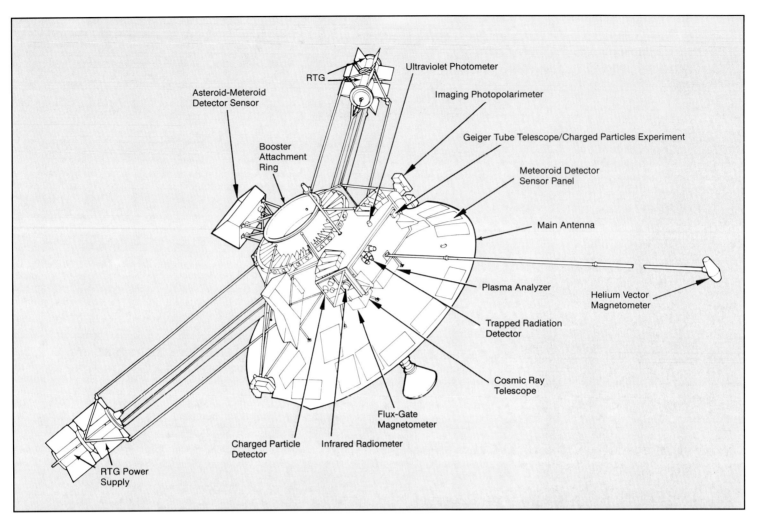

Launched 6 April 1973, **Pioneer 11** (the sister ship to Pioneer 10) came as close to Jupiter as 26,000 miles, compared to the 80,000-mile closest approach of Pioneer 10, on 4 December 1973. It provided additional detailed data and pictures on Jupiter and its satellites, including the first look at Jupiter's north and south poles which cannot be seen from Earth. This view was possible as Pioneer 11 was guided so that Jovian gravity actually threw the craft out of the plane of the ecliptic in which the planets lie. From above this plane, Pioneer 11 was able to confirm that the heliosphere extends outward in all directions from the Sun and is broken into northern and southern hemispheres by a bobbing sheet of electric current.

Pioneer 11 passed nearby Saturn on 1 September 1979, demonstrating a safe flight path for the more sophisticated Voyager spacecraft that NASA would launch in 1977. It provided the first close-up observations of Saturn, its rings, satellites, magnetic field, radiation belts and stormy atmosphere. It showed areas, smaller but resembling the Great Red Spot on Jupiter, in Saturn's clouds.

In July 1988, as Pioneer 11 passed through the outer reaches of the Solar System and Pioneer 10 sped far beyond the planets, these distant spacecraft measured drastic changes in the flow patterns of the solar wind, which is a million-mile-an-hour stream of charge particles that constantly boil off the Sun.

Considering the possibility, however remote, that Pioneer 10 may encounter intelligent extraterrestrials in the distant reaches of the universe far beyond our Solar System, NASA equipped Pioneer 10 with a plaque. The plaque has diagrams, sketches and binary numbers indicating where, when and by whom Pioneer 10 was launched.

Thanks to the Pioneers, NASA scientists have discovered a connection between the speed changes in the solar wind (near the spacecraft) and periodic changes in the Sun itself. The Sun's constant variations are manifested in shifts of its magnetic field and movements in the hot gases of its corona. Streams of faster wind particles tend to flow from thin areas, called corona holes, in the corona. Solar wind changes also are triggered by movements of a vast electromagnetic structure, called the current sheet, which bisects the Sun's field. Particles slow down as this sheet 'flaps' toward them.

Between 1985 and 1988, the Sun underwent a phase called solar minimum—a turning point in its 11-year cycle. 'No one knew what happened during solar minimum in the farthest reaches of the Solar System and beyond until the Pioneers and Voyager sent back their measurements. This is the first minimum for which we have been able to see what's going on in the solar wind out past Pluto,' recalled NASA astrophysicist **John Mihalov**.

The solar wind streams out from the Sun and envelops the entire Solar System in charged particles, mostly electrons and protons. No one knows exactly how far this five-particle-per-cubic-centimeter flow of particles extends. One recent guess is about 18 billion miles, or four times the distance of Neptune from the Sun.

Before 1985, Pioneer 10, positioned near the equatorial plane, measured periodic gusts in the solar wind called 'high speed stream.' The particles would speed up and then slow down about once every 27 days. In June 1985, the wind stream pattern stopped and the winds slowed down dramatically when Voyager 2 was two billion miles from the Sun. There was no slowing measured at Pioneer 11, about the same distance as Voyager 2, but 15 degrees higher in latitude. Pioneer 11 measured the usual pattern of high speed streams. Eventually, the winds were flowing only about half as fast at Voyager 2 as they were at Pioneer 11.

Opposite above: **Pioneer 11.** *Opposite below:* **This 1979 computer representation of Saturn's rings shows data never seen before.** *Above:* **Jupiter as seen by Pioneer 11.**

Three months later, in August 1985, the solar wind slowed and the high speed streams also stopped at Pioneer 10, which was out twice the distance of the other two probes and in the equatorial region. Mihalov believed that this change was connected to the earlier wind speed decrease at Voyager 2. The first slower particles, which blew past Voyager 2 in June, would have just reached Pioneer 10 by August. Solar winds actually sped up at the higher altitude position of Pioneer 11.

At the end of 1989, using extensive light measurements made by the Pioneer 10 and 11 spacecraft, a NASA scientist produced 'celestial constants' that will be highly useful to astronomers and physicists. The new constants were the first 'pure' measurements of the various kinds of background light in the Solar System, galaxy and universe.

This work, combined with other measurements, also provided a clue to the chemical composition of solar, galactic and cosmic dust. It gave an accurate measure of the Sun's position above the plane of the galaxy—about 12 parsecs. It described how cosmic dust scatters light. For the entire celestial sphere, 60 percent of light was scattered, not absorbed, predominantly in the same direction it had been traveling in.

Pioneer 10 spacecraft, the first to leave the Solar System, reached another milestone on 22 September 1990. At 4:00 pm EDT, Pioneer 10 was 50 times farther from the Sun than the Sun was from Earth. Reaching the 50 astronomical unit (AU) distance 'marked a new epoch in exploration of the outer Solar System,' according to Pioneer experimenter **James Van Allen (1941-)** of the University of Iowa. Van Allen, discoverer of the radiation zone around the Earth which bears his name, said reaching the 50 AU distance 'had been a goal of physicists for many decades.' When Pioneer 10 reached that mark, it was 4,647,809,899 miles from Earth, traveling farther than any manmade object.

The Viking, Venera and Magellan Projects

Because of its perceived similarity to Earth, Mars has interested and intrigued earthbound observers for centuries, and that was certainly still the case in the twentieth century. Known as the 'Red Planet' because of its distinctive iron oxide coloration, Mars reminded early observers of a distant bloody battlefield, and the Chaldeans called it Nergal, the furious one. The Greeks and the Romans named it Aries and Mars after their gods of war, and the latter name is still in use.

In 1659, the Dutch astronomer **Christiaan Huygens (1629-1695)**, who was the first to identify a Martian surface feature (Syrtis Major), also calculated the Martian day to be almost the same as the Earth's, which it is. Seven years later the Italian astronomer **Giovanni Domenico Cassini (1625-1712)** discovered the Martian ice caps. By 1783 **William Herschel (1738-1822)** had correctly calculated the exact length of the Martian day and the exact inclination of its axis.

Below: **An illustration of the Viking above Mars.** *Right top:* **Mars as seen by Viking.** *Right bottom:* **Mars as seen from Earth.**

By the middle of the nineteenth century the picture painted of Mars was that of a hospitable place that 'certainly' supported life in some form, probably similar to that of Earth. After all, their days were the same length and their seasons were parallel to ours. The darker areas on Mars were thought to represent 'vegetation,' and some observers recorded that this 'vegetation' waxed and waned with the Martian seasons.

In 1877, the Italian astronomer **Giovanni Schiaparelli (1835-1910)** had thought he'd discovered channels, or *canali*, on the surface of Mars! Translated into English as 'canals,' the features were quickly ascribed to artificial origin. It was thought that intelligent creatures had constructed an intricate system of irrigation canals on Mars to bring water from the polar ice caps to the warmer equatorial region. In 1894, the American astronomer **Percival Lowell (1855-1916)** opened his observatory at Flagstaff, Arizona primarily for the purpose of studying Mars. Lowell carefully observed and mapped the Martian surface and became a leading exponent of the idea that the canals were constructed by living creatures to irrigate their crops.

In the l930s, Eugenios Antoniadi, a Greek-born astronomer working in France, produced a map of Mars which was quite accurate for its day, but one which rejected the earlier notion of artificially constructed canals. By the late twentieth century the canal theory had been thoroughly discredited as having been an optical illusion, but the idea of Martian vegetation survived until interplanetary spacecraft from Earth visited the red planet.

The first spacecraft to pass near Mars was the American **Mariner 4** in 1965, and it was followed by **Mariner 6** and **Mariner 7**, four years later. The data returned by these flybys seemed to confirm the notion that Mars was a dull and lifeless place, roughly cratered and more like Mercury than Earth. In 1971, however, the **Mariner 9** spacecraft was placed into orbit around Mars, and the initial perception began to change. For the first time the full range of the planet's wonders, such as the great shield volcanos and the vast networks of river beds, was revealed. Mariner 9 remained in

service until October 1972, by which time the entire Martian surface had been mapped.

In August and September 1975, the United States launched the two identical Viking spacecraft toward Mars. Each Viking consisted of an orbiting module and a landing module designed to make a soft landing on the Martian surface. The Viking project was an outstanding success.

Viking 1 was launched on 20 August 1975 from Kennedy Space Center at Cape Canaveral, Florida, and inserted into Mars orbit on 19 June 1976. The 2353-pound Viking 1 Lander was detached from the orbiter and successfully landed on the Martian surface on 20 July 1976, becoming the first US spacecraft to land on another planet, from where it continued to transmit data and photographs until November 1982. The 5125-pound Viking 1 Orbiter continued to transmit data until 7 August 1980.

Viking 2 was launched from Kennedy Space Center on 9 September 1975, and inserted into Mars orbit on 7 August 1976. The Viking 2 Lander was detached and successfully landed on Mars on 3 September 1976, from where it continued to transmit data and photographs until 12 April 1980. The Viking 2 Orbiter continued to transmit data until 25 July 1978.

Each Viking spacecraft was launched by a Titan/Centaur—a Centaur upper stage combined with a Titan 3 booster. Centaur was the first United States high-energy liquid hydrogen/liquid oxygen rocket. It developed 30,000 pounds of thrust at

SIR WILLIAM HERSCHEL'S OBSERVATIONS OF MARS

William Herschel's observations of Mars began the modern physical study of that planet. He noted the expansion and contraction of white polar caps as Martian winter and summer alternated for the respective hemispheres and concluded they were snowy deposits from 'a considerable, though moderate, atmosphere.' The dark spots, on the other hand, were permanent. In 1781, he calculated the rotation period of Mars at 24 hours, 39 minutes, 21.67 seconds, which is almost precisely correct. Because of the Earth's faster orbital motion inside the orbit of Mars, that planet apparently loses one rotation during a revolution from opposition to opposition, so that an average minus correction of slightly more than two minutes has to be included.

Herschel also observed the nature of the belts on Mars, as well as irregularly drifting spots on the belts. From determinations of the variable brightness of the four Galilean satellites, he concluded that as their stellar magnitudes varied with orbital position, they rotated in the same periods as their revolutions around their primary.

Below: **The Viking spacecraft.** *Bottom:* **The Viking Lander.** *Opposite:* **Viking shown encapsulated in its payload shroud intended to protect the spacecraft during launch.**

liftoff. This was the first successful operational flight of the Titan launch vehicle with a Centaur upper stage.

The Viking project expanded humankind's knowledge of Mars manyfold and returned spectacular close-up photographs of the Martian surface that spanned Mars' seasonal changes for more than a Martian year. Viking answered a great many questions about Mars, but the notion of Martian life remains an enigma. There were three biology experiments aboard the Viking landers that were specifically designed to detect evidence of Martian life, but the answer returned was a resounding 'maybe not.' In each experiment, samples of Martian soil were scooped up by the landers' remote surface sample arms and brought aboard the spacecraft.

The **Pyrolytic Release Experiment** was designed to determine whether Martian organisms would be able to assimilate and reduce carbon monoxide or carbon dioxide as plants on Earth do. The easily monitored isotope carbon-14 was used and the results were described as 'weakly positive.' While the experiment could not be repeated by Viking on Mars, parallel experiments on Earth showed that the same results could possibly be explained by chemical, rather than biological, reactions.

In the **Labeled Release Experiment** an organic nutrient 'broth' was prepared and 'fed' to some samples of Martian soil, again using carbon-14 as the trace element. If microorganisms were present, they would 'breathe out' carbon dioxide as they 'ate' the nutrients. Carbon dioxide was, in fact, detected! However, the outgassing of carbon dioxide stopped and could not be restarted. This could have indicated some sort of chemical reaction or that a microbe *had* been present, but that it had died while 'eating' the 'broth.' To distinguish a chemical reaction from a biological reaction, the mixture was heated. This process stopped whatever it was that was producing the carbon dioxide, which *should* have ruled against the notion of a chemical reaction, but which might confirm that it had been caused by a now-deceased organism. In the end, the Labeled Release Experiment was labeled 'inconclusive' because the activities of whatever produced the carbon dioxide had no exact parallel with known reactions of Earth life.

The **Gas Exchange Experiment** was designed to examine Martian soil samples for evidence of gaseous metabolic changes, by again mixing a sample with a nutrient 'broth.' Because the Martian environment is so dry, it was decided to gradually humidify the samples before plunging them into the 'broth' so as not to 'shock' any of the life forms that might be present. A major shock came instead to the Earth-based experimenters as the sample was being humidified—there was a sudden burst of oxygen! When the nutrient 'broth' itself was added, there was some evidence of carbon dioxide but no more oxygen. Once again the results were described as 'inconclusive' because the results could not be explained by known biological reactions. Subsequent studies have been done to attempt to determine whether some type of oxidizing agent exists in the Martian soil which could provide a 'chemical reaction' explanation of the strange results of the Gas Exchange Experiment.

The three Viking biology experiments raised some curious questions, but there remain no conclusive answers to the question that has intrigued Earth-bound observers. Perhaps the answer lies closer to the Martian poles where there is more water, or perhaps the question might be restated as whether life *might have existed* at one time on Mars. In the long-gone days when rivers ran on the Martian surface, did some civili-

DISCOVERIES MADE BY THE VIKING SPACECRAFT

- The Martian atmosphere, although too thin for most living things on Earth to survive (about 1/100th as dense as Earth's), contains all components necessary to sustain life: nitrogen, carbon, oxygen and water vapor.
- The Martian surface resembles deserts on Earth but is drier than Earth's driest desert. However, considerable quantities of water are locked in the north polar ice cap and in the form of subsurface permafrost.
- Mars' northern and southern hemispheres are very different climatically. Global dust storms originate in the southern hemisphere while water vapor is comparatively abundant only in the far north during its summer.
- No canals or artifacts of any kind were found on Mars. However, evidence was found that in the past Mars' atmosphere was much thicker, its temperature was warmer and it had water on its surface.
- While volcanic mountains—at least one of which is bigger than any on Earth—were found on Mars, the planet is seismically much less active than Earth.

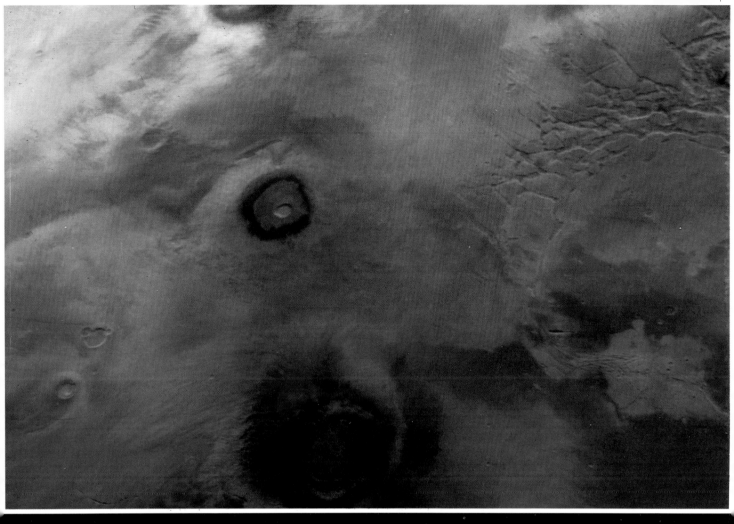

Left above: An oblique view of the red planet across the Argyre Planitia taken by Viking Orbiter 1. This large impact basin is visible from Earth. The clouds on the horizon are thought to be crystals of frozen carbon dioxide. *Left:* The Viking mission far exceeded expectations in regard to scientific data returned from Mars. Parts of the Viking Lander are visible in this image of Mars. *Above:* Three large Martian volcanoes on the Tharsis Ridge: Arsia Mons, Pavonis Mons and Ascraeus Mons. *Right:* The Martian surface at the Utopia Planitia landing site.

zation, or even just a species of moss, flourish on their banks? Will paleontologists or archaeologists from Earth one day discover fossils or ruins amid the drifting rust-red Martian sands?

As viewed from Earth, Venus is the brightest celestial object in the sky except for the Sun and Moon. Like the Moon, Venus can be seen to go through a series of phases as it orbits the Sun and is viewed from Earth. The Greek poet Homer even went so far as to call it the most beautiful star in the sky, while the Romans named it Venus after their goddess of beauty. The second planet from the Sun, Venus is a near twin of the Earth in terms of size, with a diameter 95 percent that of our own planet. Like Mercury, and unlike every other planet in the Solar System, Venus has no moon.

By the early twentieth century, it had been determined that the Venusian surface was obscured by clouds, and various theories evolved regarding the actual nature of the surface beneath those clouds. The nineteenth century idea that the planet was covered by lush jungles was dismissed in favor of the two schools of thought that suggested either a vast desert or a vast ocean of water.

It had been established that the surface would be extremely hot because carbon dioxide in the thick atmosphere would prevent solar heat from escaping the surface, thus producing what is referred to as a 'greenhouse effect.'

The first successful expedition to the vicinity of Venus came in December 1962 when the American unmanned spacecraft **Mariner 2** traveled to within 21,600 miles of the

planet. The flight of Mariner 2 was a major milestone in unlocking the secrets of the mysterious planet. Among its achievements were confirmation that Venus has no detectable magnetic field, confirmation of the planet's exact rotational period—245 Earth days—and confirmation that it rotates from east to west, rather than the opposite as previously supposed.

Mariner 2 also provided a more accurate reading of the planet's surface temperature, which at 900 degrees Fahrenheit is too hot for the existence of an ocean, because water could exist there only as steam. Water vapor is, in fact, present in the atmosphere, and some astronomers have theorized that at an early stage in the evolution of Venus, oceans *may have*, in fact, existed on the surface.

In 1978, the United States undertook the **Pioneer Venus** project as a follow-on to several earlier Mariner probes. The project consisted of an orbiter spacecraft and a multiprobe spacecraft. The former undertook the detailed radar mapping of the Venusian surface that made possible the maps on these pages, and which gave us much of the information we now have about the planet's terrain. The multiprobe was actually five probes designed to return data about the Venusian atmosphere as they plunged toward the surface. One of the Pioneer Venus multiprobes continued to return data from the surface for just over an hour after impact.

Opposite: **Four views of Venus taken by the Pioneer Venus Orbiter.** *Below:* **The interaction of the solar wind with the Venusian atmosphere.**

The Soviet Union, meanwhile, prepared a series of spacecraft to conduct soft landings on the Venusian surface, which returned the only photographs ever taken of the Venusian surface. The Soviet **Venera 9** and **Venera 10** spacecraft each returned a single black and white image in 1975, and the **Venera 13** and **Venera 14** spacecraft returned color photos in March 1982.

While the Soviet Venera spacecraft provided the first photographs of specific points on the Venusian surface, the American Pioneer Venus Orbiter provided our first clear look at the overall global surface features of Venus.

Using a radar altimeter, Pioneer Venus was able to obtain the data necessary to produce a topographical map of 90 percent of the planet's surface, from 73 degrees north latitude to 63 degrees south latitude.

On 4 May 1989, the United States used the Space Shuttle Orbiter **Atlantis** to launch the four-ton **Magellan** radar-mapping spacecraft on a mission to Venus. Arriving at the cloud-shrouded planet on 17 August 1990, Magellan initially conducted 1852 mapping swaths around the planet, a process which continued for 243 Earth days, or one Venusian day. The resolution of the data returned by Magellan was vastly superior to that achieved by Pioneer Venus a decade earlier. Indeed, the quality of the imagery was so good that the data it transmitted looked like actual photography of a cloudless planet!

Satellite data from Pioneer Venus in 1978 and 1979 was thought to show that the surface was generally smoother than those of the other three terrestrial planets, but in 1990,

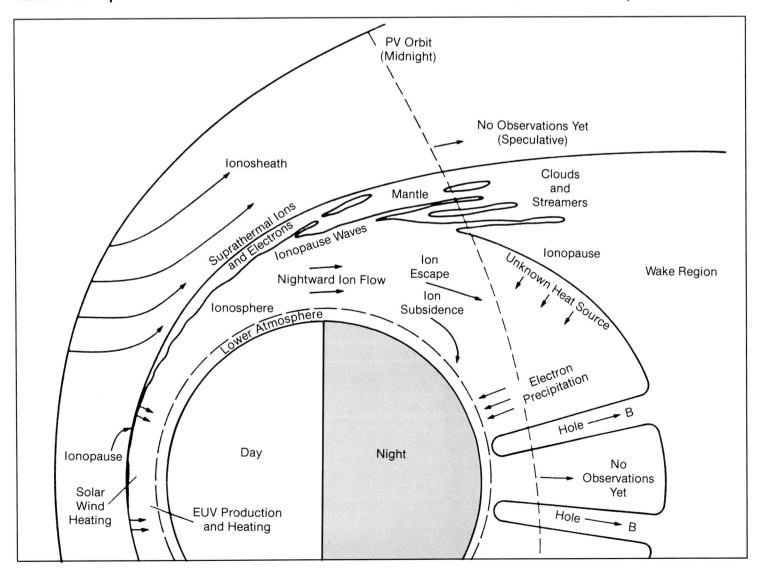

Magellan, which was capable of 'seeing' in more detail, revealed a more rugged terrain. However, Magellan confirmed Pioneer's findings that Venus has much less variation in altitude than is seen on Earth. For instance, 60 percent of the Venusian surface is within 1600 feet of the planet's mean radius of 3752 miles. It has been suggested that this is due to the deeper lowlands having been filled with sand and other wind-blown material. Because there are no seas on Venus, the mean radius is used as a reference point in the same way that sea level is used on Earth.

Most of the surface of Venus is characterized as rolling uplands, rising to an altitude of roughly 3000 feet, while 20 percent of the surface is identified as lowlands and 10 percent as mountainous. The two largest upland regions, or continental masses, are Aphrodite Terra (roughly the size of Africa), near the equator in the Southern hemisphere, and Ishtar Terra (roughly the size of Australia), in the northern hemisphere near the North Pole. These two features constitute the Venusian 'continents' and are named respectively for the ancient Greek and ancient Babylonian goddesses of love.

The highest points on the mapped surface of Venus are in the Maxwell Mountains (Maxwell Montes) in Ishtar Terra. High enough to have been identified by Earth-based radar prior to the Pioneer Venus project, the Maxwell Mountains, which may actually be a single mountain, rise to more than 35,000 feet above mean radius, or roughly 20 percent higher than Mount Everest rises above Earth's *sea level*. If viewed from the surface they would be an impressive sight, rising nearly 27,000 feet above Lakshmi Planvin, the surrounding plateau which is roughly the same elevation as the Tibetan plateau on Earth.

Data obtained from Pioneer Venus indicates that the Maxwell Mountains may be the rim of an ancient volcano whose caldera had a diameter of roughly 60 miles. The lava flows, however, have long since been worn away by wind erosion, and the slopes of the Maxwell Mountains are strewn with rocks and debris.

Another important upland region is Beta Regio with its great shield volcanos, Rhea Mons and Theia Mons, which are larger than the great shield volcanos of Hawaii on Earth. The mountainous Beta Regio is still in the process of formation and probably contains active volcanos. As such, it is the newest major surface feature on Venus.

The lowest point on the Venusian surface is actually a canyon, Diana Chasma, located within central Aphrodite Terra. At just 9500 feet below mean radius, Diana Chasma is much shallower than the corresponding lowest point on Earth, the Marianas Trench. The largest and lowest lowland region on Venus is the Atalanta Plain (Atalanta Planitia) located northeast of Aphrodite Terra and due east of Ishtar Terra. It is roughly the same size as the Earth's North Atlantic Ocean, although it is shallower by comparison.

On the surface of Venus, atmospheric pressure is roughly 90 times that of the Earth. A yellowish glow like that of a smoggy sunset on Earth is all pervasive. The daytime surface temperature of 900 degrees Fahrenheit is a global constant because the carbon dioxide and sulfuric acid atmosphere and cloud cover function as an insulating blanket, trapping the heat and producing convection currents that redistribute it across the entire surface area.

By January 1991, the Magellan spacecraft, mapping the surface of Venus with imaging radar, had swept over nearly 55 percent of the planet, an area comparable on Earth to the distance from Los Angeles to New Delhi, India. Scientists

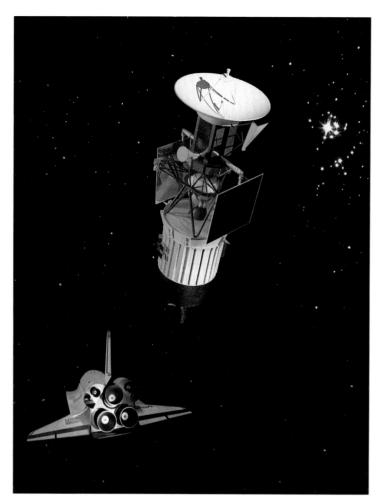

Opposite: **Testing the Magellan spacecraft.** *Above:* **Magellan attached to the Inertial Upper Stage (IUS) rocket.** *Below:* **An artist's drawing of Magellan as it begins its orbit of Venus.**

Above: **A Magellan image of the eastern edge of Lakshmi Planum and the western edge of Maxwell Montes (upper left). *Opposite left:* The continent Ishtar and outlined in yellow, Aphrodite; also seen *(opposite right)*. *Right:* Venusian highlands.**

reported that the radar mapping data provided them with significant new knowledge about the surface of Venus and its atmosphere.

According to Project Scientist **Dr Steve Saunders**, 'All of the areas mapped showed widespread evidence of volcanism along with evidence of tectonics, the process that produces mountains. Venus and Earth are the only planets in our Solar System that have linear mountain belts. However, the mountains on Venus are not deeply eroded by rainfall and running water as are mountains on Earth.'

Magellan also confirmed the number of Venusian impact craters that scientists had expected to find, judging from their earlier Earth-based radar data. The smallest impact craters were about three miles in diameter, indicating the dense Venusian atmosphere had effectively shielded the surface from bombardment by smaller asteroids and comets.

'There also was evidence that the poisonous, thick atmosphere of Venus was not formed recently,' Saunders said. 'Surface images indicated it may be from 400 million to 800 million years old, or even older. The Venus atmosphere is 90 times heavier than that of Earth, and composed primarily of carbon dioxide, with significant amounts of sulfuric acid at high levels.'

New styles of volcanism were found, and lava channels, hundreds of miles long, were noted at several places on the plains. Although lava channels have been found on Earth, none were as long or as regular as those seen on Venus. Another new type of volcanism, referred to as 'pancake' domes by scientists, formed structures on the plains up to 20 miles across and nearly a mile high. Scientists thought the domes were formed by an outflow of a pasty, thick lava, similar to silicon-rich lavas on Earth. Volcanic domes also form on Earth, but are much smaller and form in volcanic calderas.

The images sent back by Magellan also revealed indications of turbulent surface winds on Venus. The evidence was in the form of wind streaks in the lee of topographical obstacles, such as the small, low shield volcanos on the plains.

As least four competing theories about the nature of Aphrodite Terra, a continent-sized highland, were being tested by the new Magellan data. Aphrodite Terra, the largest highland region on Venus, extends nearly two-thirds of the way around the planet. Scientists were able to study radar images of the western portion of Aphrodite, called Ovda Regio. Earlier data produced by the Pioneer Venus Radar Mapper indicated that the topography in this region might be similar to Earth's continents. The various theories were based on the Pioneer data and other earlier radar imagery, as well as topography and gravity data.

The first hypothesis was the 'ancient continent' model, which held that Aphrodite formed from lighter rock that crystallized early and literally floated on the dense mantle of Venus. There are similar formations on Earth and the Moon. A second theory, called the 'spreading ridge' model, compared the topography of Aphrodite to that of mid-ocean ridges on Earth where new crust is being formed as the continents drift apart. Another hypothesis advocated the 'hot

THE VIKING, VENERA AND MAGELLAN PROJECTS 113

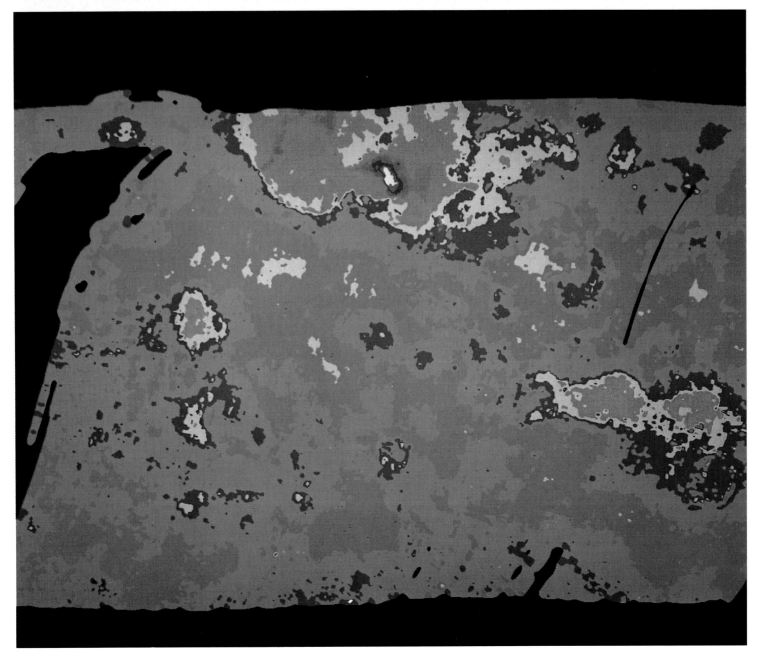

spot' model, which states that the equatorial highlands were pushed up by hot spots similar to the ones that lie beneath the island of Hawaii. Hot spots are regions of mantle that tend to rise, producing broad domes that frequently result in huge volcanos. The fourth hypothesis was that western Aphrodite was a region of 'mantle downwelling.' That means a downwelling plume of colder, more dense mantle material caused the surface crust to compress and thicken. The thicker region would then stand higher than its surroundings.

'The radar imager used to penetrate Venus' thick clouds and map the surface worked flawlessly,' said Project Manager **Tony Spear**. He added that the spacecraft was sufficiently healthy to continue mapping Venus well into the 1990s, 'providing a windfall of additional, exciting science return. The first objective of the extended Magellan mission was to collect the remaining 16 percent of the map, including the south pole, which had never been imaged. Other surface features were imaged from a different view angle to yield a new perspective, and image comparisons from one mapping cycle to the next were made to look for surface activity.'

Widespread volcanism and a geologically active surface were descriptions of the planet Venus presented on 12 April 1991 in the first papers published in *Science* magazine by members of the Magellan Project science team at NASA's Jet Propulsion Laboratory in Pasadena, California. The papers described geologic features of Earth's sister planet as scientists began the unprecedented task of mapping an Earth-size planet. Their reports described extensive and explosive volcanism, tectonic deformations, mountain belts and a number of impact craters that indicated Venus has a relatively young surface age of a few hundred million years. Science team members described several types of lava flows, evidence of lava rivers hundreds of miles long and craters created by meteorite impacts that caused surface material to be ejected as far as 600 miles into the air.

By the end of July 1991, Magellan had circled Venus 2351 times and had traveled 75 million miles around the planet in its mapping mission. One complete circuit of Venus took Magellan eight months, the length of one Venus day. When Magellan finished its mapping mission, it had gathered more than double the amount of all other image data collected to date in the US planetary program.

In August 1991, Magellan discovered the longest channel known in the Solar System. 'The channel crosses the plains of Venus for 4200 miles—longer than the Nile River, the longest river on Earth,' said Saunders. 'The very existence of such a long channel is a great puzzle. If the long channel were carved by something flowing on the surface, the liquid must have had unusual properties. There are no very likely candidates for a liquid. Lava, even very high temperature types, would need to have a very high extrusion rate to flow so far. This is not consistent with the uniform narrow channel morphology.'

Saunders added it may have been some material that was near its freezing or melting point at the average surface conditions of Venus—surface pressure 90 times that of Earth and a temperature of 864 degrees Fahrenheit.

Left: **The Navka region of Venus is a volcanic plains region.**
Right: **A segment of a sinuous channel on Venus.**

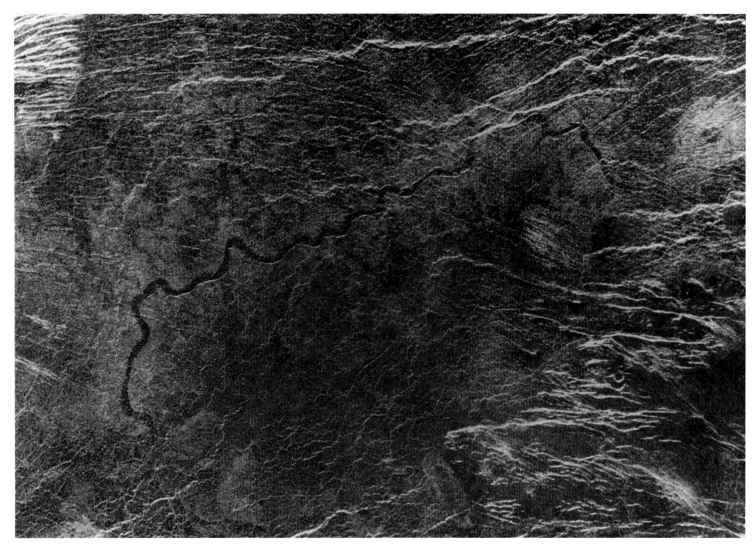

The Voyager Project

By 1970, NASA seemed poised on the threshold of a fabulous epoch in Solar System exploration. The Apollo program had sent four manned spacecraft into lunar orbit, and two pairs of American astronauts had actually walked on the Moon's surface. Within two years, a dozen men would tread on lunar soil. A manned mission to Mars was planned for 1981.

Simultaneously, serious interest in the idea of sending an unmanned spacecraft on a **Grand Tour** to the Solar System's outer planets was aroused within the American aerospace community. Led by enthusiasts at the Jet Propulsion Laboratory, work started on planning for the 1977 **Jupiter-Saturn-Pluto** mission and a 1979 **Jupiter-Uranus-Neptune** mission. At one time, these were to be launched on a Saturn 5 launch vehicle (JPL) (based on the Saturn 5 that sent Apollo to the Moon) and utilize a new **Thermoelectric Outer Planets Spacecraft (TOPS)** equipped with a Self-Test-and-Repair (STAR) computer. Industrial giants, such as Rockwell, Boeing, Martin Marietta and TRW, invested significant company funds on large, in-house study teams to prepare themselves for the Grand Tour project bidding. The 17th Annual Meeting of the American Astronautical Society in June 1971 at Seattle was entirely devoted to the Grand Tour and Outer Solar System Missions. The success of the Apollo Lunar Missions caused a flare-up of United States public interest, only to be followed by the hangover of an expensive victory in the unopposed Moon race, and the project cost of the TOPS/JSP/JUN mission soared to about one billion dollars. The legislative and public apathy that resulted caused a funding demise, the effects of which are felt to this day. As a consequence, funding for the Grand Tour was not included in the 1972 budget, the effort was officially canceled and the industrial teams were disbanded or laid off.

Promising ideas do not die easily, however. The 179-year reoccurrence cycle of an outer planets Grand Tour opportunity made the 1977 launch too precious an asset to bypass, because another such launch 'window' would not occur until the year 2156 AD.

A Mariner-type, low cost, JPL in-house mission to Jupiter and Saturn was approved by mid-1972 through the efforts of steadfast space science advocates in Congress, NASA and at JPL. The twin spacecraft were tentatively called **Mariner 11** and **12**, both to be launched in August-September 1977 on Titan 3D launch vehicles. The cost of the new **Mariner-Jupiter-Saturn** 1977 project was not to exceed 360 million dollars. The two-spacecraft mission was scheduled to be a reconnaissance effort at Jupiter that would emphasize Io and Ganymede. At Saturn, intensive investigations by both spacecraft of Titan and Saturn's rings were also contemplated.

By 1975, the astounding radiation belt intensities at Jupiter were sampled by the **Pioneer 10** and **Pioneer 11**, while another planned **Mariner-Jupiter-Uranus** mission in 1979 was experiencing funding difficulties. The latter project was then redirected to become **Project Voyager**. One spacecraft would go to Io at Jupiter and the other to Titan at Saturn. Mission designers retained the option of going to Uranus, and this, of course, meant a green light to Neptune as well. The Grand Tour still had a chance to be accomplished—with all its promise of discovery—in a single shot and only 12 years of flight. In March 1977, the project and the spacecraft were officially renamed **Voyager 1** and **Voyager 2**.

Voyager 1 was launched from Cape Canaveral on 5 September 1977 and targeted toward Jupiter, which it encountered at a distance of 177,720 miles, on 5 March 1979. A gravity-assist from Jupiter slung the spacecraft on to its next destination, Saturn, which it observed from as close as 77,000

Below: Voyager 2 sees the Sun rise over Jupiter. *Opposite:* Voyager 2 discovered two more rings in Saturn's complex system.

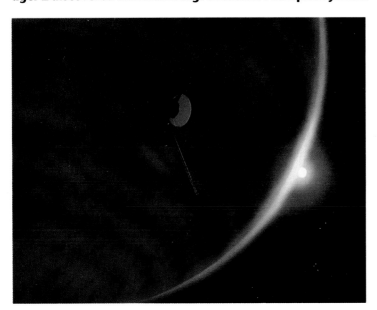

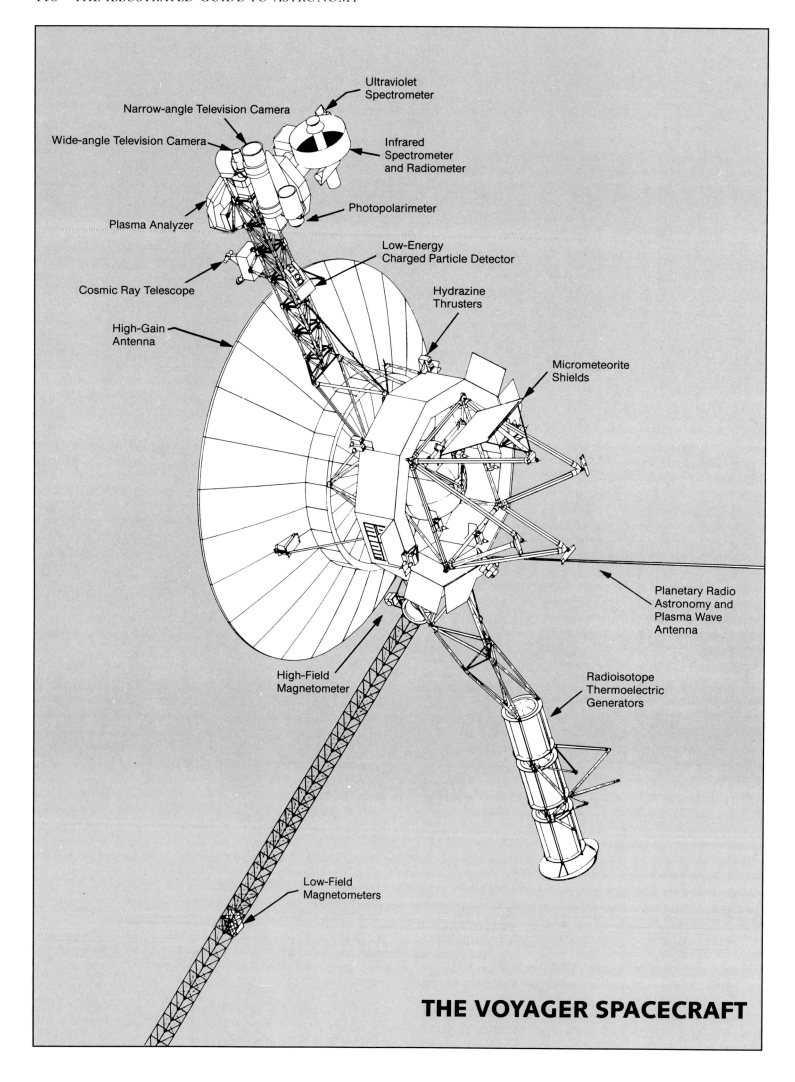

miles on 12 November 1980. Mission objectives at Saturn, including a very close flyby of the giant moon Titan, precluded Voyager 1 from continuing on to any other planets.

Voyager 2 was launched on 20 August 1977, about two weeks before the launch of Voyager 1. Flying a longer trajectory, it reached Jupiter, coming as close as 399,560 miles, on 9 July 1979, four months after Voyager 1's flyby. Like its twin spacecraft, Voyager 2 was slung on to Saturn, which it encountered at a distance of 63,000 miles on 25 August 1981. At that time, Voyager 1 had been directed north, past Saturn's moon Titan, while the decision was made to send Voyager 2 to an aimpoint at Saturn, which would permit it to follow a course toward Uranus that allowed it to pass within a half million miles of that planet in 1986 and continue outward in the Solar System toward its encounter with Neptune in 1989.

The spacecraft themselves weighed 1819 pounds, including a 258-pound instrument payload. Like the Mariner spacecraft and Viking Mars Orbiters, the Voyagers are stabilized on three axes, using the Sun and the star Canopus—one of the brightest in the galaxy—as celestial references.

The dominant feature of these spacecraft is their 12-foot diameter, high-gain antenna that is pointed almost continuously at Earth. While the antenna dish is white, most visible parts of the spacecraft are black—blanketed, or wrapped, for thermal control and micrometeoroid protection. A few small areas are finished in gold foil or have polished aluminum surfaces to reflect heat.

Data storage capacity on the spacecraft's tape recorders is about 536 million bits—approximately the equivalent of 100 full-resolution photos. Dual-frequency communications lines, such as S-band and X-band, provide accurate navigation data and large amounts of science information during planetary encounter periods—up to 115,200 bits per second (bps) at Jupiter, 44,800 bps at Saturn and 21,600 bps at Uranus.

Voyager's discoveries were numerous. They included a ring of rocky debris and an ionized torus of sulfur around Jupiter, and vast thunderstorms, sometimes big enough to engulf Earth, on Jupiter and Saturn. In addition, it discovered many additional small satellites, the natures of the Galilean satellites and Saturn's largest satellites, discoveries about Saturn's rings such as structure, width, unusual braiding, spoke patterns, ringlet formations, additional rings and numerous ringlets filling most of what appear from Earth to be gaps in Saturn's rings.

Voyager's close looks at Jupiter's satellites showed that gaudy, sulfurous Io is the most volcanic object in the Solar System; Europa has the smoothest surface; Ganymede had some tectonic activity before it froze solid about three million years ago; and Callisto's crater-pocked surface is as ancient as that of the Moon or Mercury. Ganymede and Callisto are composed of about equal parts of water, ice and rock; Europa has substantial quantities of water, while Io is a waterless moon.

Voyager's findings indicate that Saturn's rings and satellites are, for the most part, composed of water ice. The exceptions are the giant satellite Titan, which is half rock and the outermost satellite, Phoebe, which is mostly rock. Voyager photographs of Phoebe show that it resembles asteroids. It may be a captured outer Solar System asteroid. Voyager measurements also toppled Titan from its position as the Solar System's largest satellite. They found Jupiter's Ganymede to be slightly larger. Both are bigger than either of the planets Mercury or Pluto.

Opposite: **A three-quarter view of the Voyager spacecraft.**
Above: **The Voyager spacecraft under construction.**

Voyager 1 discovered that nitrogen makes up about 82 percent of Titan's atmosphere. (Nitrogen is 78 percent of Earth's atmosphere.) The other known ingredients of Titan's atmosphere are methane, ethane, acetylene, ethylene, hydrogen, cyanide and other organic compounds. Titan's atmosphere is believed to be similar to Earth's in primeval times. Titan's composition reflects an abundance of water. An orange-colored smog, thought to result from the chemical reaction of solar radiation on the methane in Titan's atmosphere, envelops the satellite, preventing direct observation of its surface. Voyager found Titan's surface air pressure to be 1.6 times that of Earth's sea level pressure. Some scientists speculated that the smog in Titan's atmosphere could have created a greenhouse effect that over the last few billion years warmed the planet enough for organic chemicals to evolve toward prelife forms. Voyager 2 measurements show that Titan's surface temperature is 95 degrees Kelvin (-288 degrees Fahrenheit), too cold for water to liquefy or for significant progress in prelife chemistry.

Around this temperature, methane could exist as a liquid, vapor or solid, just as water does on Earth. On Titan then, methane rain or snow may fall from methane clouds. Methane rivers may flow through icy methane channels and methane oceans may fill icy basins.

The Voyagers also confirmed that neither Jupiter nor Saturn have solid surfaces. Their data indicated that while Jupiter's radiant heat may originate from gravitational contrac-

120 THE ILLUSTRATED GUIDE TO ASTRONOMY

THE VOYAGER PROJECT

Below: The Voyager spacecraft, shown without thermal blankets for clarity, is a compact investigative research unit weighing 1819 pounds. The gold emblem seen plainly here, is a map to Earth, the origin of the spacecraft. It is also equipped with a recording of 'Sounds of Earth' that includes natural and manmade sounds.

tion or release of heat accumulated during its formation, Saturn's comes from gravitational separation of helium and hydrogen in Saturn's interior. We now know that both planets radiate about 2.5 times more heat than they receive from the Sun.

Before the flyby of Voyager 2 in January 1986, Uranus was thought not to have a magnetic field, but this assumption proved false. The magnetic field of Uranus is tilted at a 60 degree angle to the planet's rotational axis (compared to 12 degrees on Earth). The magnetic field has roughly the same intensity as the Earth's, but whereas the Earth's magnetic field is generated by a molten metallic core, the one surrounding Uranus seems to be generated by the electrically conductive, super-pressurized ocean of ammonia and water that exists beneath the atmosphere.

Prior to 1989, very little was known about Neptune because of its great distance—nearly three billion miles—from Earth. Voyager 2 turned its cameras on Neptune in June 1989 and flew to within 3044 miles of the planet on 25 August. During this short time, mankind's knowledge of Neptune and its moons increased one hundred fold. For example, six moons and four rings were discovered, and the existence was confirmed of a magnetic field tilted 50 degrees from Neptune's rotational axis and offset 6000 miles from the planet's center. The temperature of Neptune was determined to be -353 degrees Fahrenheit.

Most notable visually was the discovery of the Great Dark Spot in Neptune's atmosphere. Located at a mean latitude of 22 degrees south and with an overall length of 30 degrees longitude, it is very much analogous to Jupiter's Great Red Spot, both in terms of location and its size relative to the planet. Like the Great Red Spot, the Great Dark Spot is an elliptical, stormlike feature that rotates in a counterclockwise direction and is probably located *above* the surrounding cloud tops. The Great Dark Spot is encircled by a constantly changing pattern of cirrus clouds and followed in its movement through Neptune's atmosphere by a string of much smaller elliptical storms. The cirrus clouds are composed of methane crystals and cling to the Great Dark Spot the way cirrus clouds cling to mountain tops on the Earth's islands. The Great Dark Spot completes a revolution of the planet in 18.3 hours, moving east to west.

To the south of the Great Dark Spot is another bright cloud which was observed by Voyager for the entire duration of the flyby and which was nicknamed 'Scooter' because it moves at a much faster relative speed than the Great Dark Spot, causing it to overtake the latter every five days. South of this feature is the Lesser Dark Spot, with a permanent cloud bank situated over its center.

Because Jupiter, Saturn and Uranus are all encircled by rings of rocky debris, it was long supposed of Neptune as well, although the planet is too distant for these to be visible from Earth. It was not until two weeks prior to Voyager 2's closest encounter, in August 1989, that the existence of these rings was verified. Because Neptune's rings are very irregular, they appeared at first to be incomplete arcs. It is now known that four continuous, albeit thin, rings encircle the planet.

Prior to the Voyager 2 encounter, the planet was known to possess two moons, although 1981 observations at the University of Arizona led to a prediction of at least one other moon. Prior to Voyager 2's encounter with Uranus in 1986, only five Uranian moons were known, and Voyager's observations tripled that number. This fact alone would lead us to suspect that Neptune, too, had moons awaiting discovery.

Above: **The Voyager 1 liftoff in 1977.** *Opposite:* **An artist's conception of Voyager's trajectory as it is flung past Saturn.**

The two Neptunian moons confirmed prior to 1989 are among the most peculiar in the Solar System. Triton, discovered in 1846 less than a month after Neptune, is a huge object with a retrograde orbit, while Nereid, discovered more than a century after Triton, has the most elliptical orbit of any known moon in the Solar System. Voyager 2 discovered six additional moons, including the one known as Proteus, which is actually larger, although darker, than Nereid. This moon is also the largest known nonsymmetrical body in the Solar System.

Triton, which revolves around Neptune in a direction opposite to the mother planet's rotation, was thought to have a diameter of 3700 miles, but which is now known to be 1690 miles. Triton is the second moon in the Solar System (after Saturn's Titan) that has been found to possess an atmosphere. It is a much thinner atmosphere than Titan's but is nevertheless clearly discernible in Voyager 2 photos of Triton's horizon. The atmosphere consists of methane chilled to -400 degrees Fahrenheit, making Triton the coldest object yet observed in the Solar System. Triton consists primarily of silicate rock, but there is also a great deal of water ice and frozen methane present.

Triton has a very pronounced division between that part of the moon which is experiencing summer and that part which is enduring winter. This phenomenon is not yet explained, nor is the fact that within the 10 years prior to Voyager's flyby, Triton was twice as orange as it appeared both from Earth and from Voyager in August 1989. Evidence of recent volcanic activity on Triton has also been discovered, so that it is now possible to add it to the short list of volcanically active bodies which previously included only the Earth and Jupiter's moon Io.

Like its parent planet, Triton is named for a god of the sea—in this case, the merman son of the Greek god Poseidon and goddess Amphitrite. Its orbit is so close to Neptune, and is gradually getting so much closer, that one day Neptune's gravity might pull it apart and scatter it into a Saturn-like ring. (However, if this does happen, it will not be for several million years.)

Nereid, Neptune's second moon, was discovered in 1949 by the Dutch-American astronomer, **Gerard Peter Kuiper (1906-1973)** in an elliptical orbit far beyond the orbit of Triton. Named for the Nereids—sea nymph daughters of the Greek god Nereis—Nereid is much tinier than its brother Triton. Little is known about Nereid other than its extremely *elliptical* orbit and its size relative to Triton. After the Voyager spacecraft flew by the four giant outer planets—Jupiter, Saturn, Uranus and Neptune—their mission might have been over. However, they were just beginning a new phase of their journey, called the **Voyager Interstellar Mission (VIM)**.

As the Voyagers cruised gracefully in the solar wind in 1990, their fields, particles and waves instruments were studying the space around them while searching for the elusive outer edge of our Solar System—the heliopause. The **heliopause** is the outermost boundary of our solar wind, where the interstellar medium restricts the outward flow of the solar wind and confines it within a magnetic bubble called the heliosphere. The solar wind is made up of electrically charged atomic particles, composed primarily of ionized hydrogen, that stream outward from the Sun.

'The termination shock is the first signal that we were approaching the heliopause,' said Voyager Project Scientist and JPL Director **Dr Edward C Stone**. 'It's the area where the solar wind starts slowing down.' Mission scientists forecast that the Voyager may cross the termination shock by the end of the twentieth century.

The exact location of the heliopause remains a mystery. It has long been thought to be some seven to 14 billion miles from the Sun. Any speculation about where the heliopause is, or what it's like, has come only from computer models and theories. 'Voyager 1 is likely to return the first direct evidence from the heliopause and what lies beyond it,' Stone predicts.

Now that Voyagers' primary mission of exploring the outer planets has been completed, there are fewer constraints on the science team when it comes to programming the spacecrafts' observations. 'We can sit on these things for very long periods of time and watch these stars go through their phases,' states Dr Jay Holberg, a member of Voyager's ultraviolet subsystem team.

'In exploring the four outer planets, the Voyagers already have had an epic journey of discovery. Even so, their journey is less than half over, with more discoveries awaiting the first contact with interstellar space,' Stone said. 'The Voyagers revealed how limited our imaginations really were about our Solar System, and I expect that as they continue toward interstellar space, they will again surprise us with unimagined discoveries of this never-before-visited place which awaits us beyond our planetary neighborhood.'

A PORTRAIT OF JUPITER

When the Solar System was being formed 4.6 billion years ago, Jupiter may have had the makings of becoming a star. At that time it was 10 times its present diameter and heated by gravitational contraction. It may have blazed like a second Sun.

Had the nuclear reactions within Jupiter become self sustaining as they did in the Sun, the two objects may have become a double star of the type that exist elsewhere in the galaxy and the Solar System would have been a vastly different place than it is today. However, Jupiter failed as a star and gradually began to cool and to collapse to its present size. As Jupiter cooled it became less brilliant, so that after a million years the 'star that might have been' went from one hundred thousandth the luminosity of the Sun to one ten billionth. Today, as much energy is still radiated from *within Jupiter itself* as it *receives from the Sun*.

Named for the king of all the Roman gods, Jupiter is the largest planet in the Solar System, and second in mass only to the Sun. This 'king' of planets has 1330 times the volume of the Earth and 318 times the mass.

Like the Sun, Jupiter is composed almost entirely of hydrogen and helium and, unlike the terrestrial planets, it may be composed almost entirely of gases and fluids with no solid surface. If there is a solid, rocky surface, the solid diameter of Jupiter may actually be just slightly larger than the Earth's. Above this rocky surface, if it exists, there may be a layer of ice more than 4000 miles thick, which is kept frozen by *pressure* rather than temperature, as the *temperatures* would be frightfully hot.

Jupiter's magnetic field is more than 4000 times greater than that of the Earth, leading observers to postulate the existence of a metallic core. Strangely, the center of Jupiter's magnetic field is not located at the planet's center, but at a point 6200 miles offset from the center and tipped by 11 degrees from the rotational axis. This point is also the center of Jupiter's vast magnetosphere, which is six million miles across. Three million miles from Jupiter the plasma reaches the hottest temperatures recorded in the Solar System, roughly 17 times as hot as the interior of the Sun.

There is almost certainly a sea of liquid metallic hydrogen that makes up the bulk of Jupiter. Above the liquid metallic hydrogen is a transition zone leading to a layer of fluid molecular hydrogen. Jupiter's atmosphere is characterized by colorful swirling clouds that cover the planet completely. These clouds form in Jupiter's troposphere, at the altitude where convection takes place. The lower clouds, like those of the Earth, are thought to be composed of water vapor, with ice crystals being present at higher altitudes. Above these, higher clouds are composed of ammonium hydrosulfide, with Jupiter's high cirrus composed of ammonia. At a point 40 miles above the ammonia cirrus, where the Jovian troposphere gives way to the stratosphere, temperatures can dip to colder than -150 degrees Fahrenheit. Above that, in the ionosphere, however, temperatures increase again.

Jupiter's atmosphere is a complex and dynamic feature characterized by distinct horizontal 'belts,' or darker bands of clouds, that exist at semi-symmetrical intervals in the northern and southern hemispheres, and which alternate with lighter colored 'zones.'

The most outstanding feature on Jupiter is certainly the **Great Red Spot**. First observed in 1664 by the astronomer **Robert Hooke (1635-1703)**, it is a brick red cloud three times the size of the Earth. Described as a high pressure
(continued on page 126)

Left: **With the Earth shown at the same scale as Jupiter in this composite photo, the cloud features on Jupiter are apparently as large as Earth's continents. The GRS is three times the size of the Earth.** *Right:* **Jupiter as seen by Voyager. The stars in the background were added by an artist.**

THE VOYAGER PROJECT 125

system, the Great Red Spot resembles a storm and exists at a higher and colder altitude than most of Jupiter's cloud cover, although traces of ammonia cirrus are occasionally observed above it. It rotates in a counterclockwise direction, making a complete rotation every six Earth days, and it varies slightly in latitude. The Great Red Spot is almost certainly the top of some sort of high altitude updraft plume from below the Jovian cloud cover, but the exact nature of the Great Red Spot is uncertain. One theory is that it is above an updraft in the Jovian atmospheres in which phosphene, a hydrogen-phosphorus compound, rises to high altitudes—where it is broken down into hydrogen and red phosphorus-4 by solar ultraviolet radiation. The pure phosphorus would give the Great Red Spot its characteristic color. Another theory has the Great Red Spot as the top of a column of stagnant air that exists above a topographical surface feature far below, within Jupiter.

The Great Red Spot may be an awesome feature, but it is a transient one. Any storm that has been raging for more than 300 years can certainly be termed as an impressive meteorological phenomenon, but it hasn't been constant in its intensity. Between 1878 and 1882 it was seen as very prominent, but thereafter it dimmed markedly until 1891. Since then, it has waned slightly several times—in 1928, 1938 and again in 1977.

Other intriguing meteorological phenomena have also been observed in the Jovian atmosphere, including smaller red spots in the northern hemisphere and some dark brown features that formed at the same latitude as the Great Red Spot. Designated as the **South Tropical Disturbance**, these features were first observed in 1900, overtook and 'leaped' past the Great Red Spot several times, and gradually began to fade in 1935, disappearing five years later. In 1939, a group of large white spots formed near the Great Red Spot in the southern hemisphere. Like their larger red counterpart, they rotate counterclockwise. Similar, but smaller, features have been observed in the northern hemisphere, where they are seen to rotate in a clockwise direction.

Jupiter has a distinct ring system that was unknown before the close-up observations by two American Voyager spacecraft in 1979. Unlike Saturn's rings, Jupiter's rings are very thin and narrow, and are not visible except when viewed from behind the night side of the planet, when they would be backlighted by the Sun. The ring system is divided into two parts that begin 29,000 miles above Jupiter's cloud tops, although some traces of ring material exist below that altitude. The two parts are a faint band 3100 miles across, feathering into a brighter band 500 miles across. The rings are composed of dark grains of sand and dust and are probably not more than a mile thick.

Jupiter's 16 known moons are organized into a very orderly system of *four* dissimilar groups, each comprised of four similar sized moons orbiting in distinctly different planes.

The inner group of moons (except for Amalthea) was discovered during the Voyager project, and all have diameters of less than 200 miles. They all orbit in a plane whose orbital inclination is less than half a degree and they are located less than 140,000 miles from Jupiter. The second group, called 'The Galileans,' was discovered in 1610 by **Galileo Galilei (1564-1642)** and all have diameters greater than 1900 miles. They all orbit in a plane whose orbital inclination is less than

(continued on page 129)

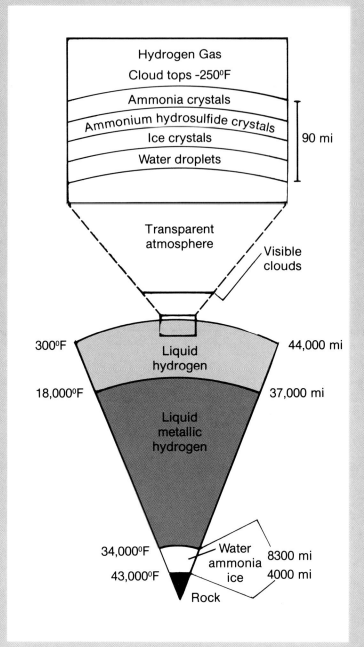

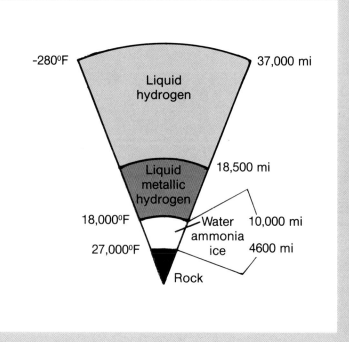

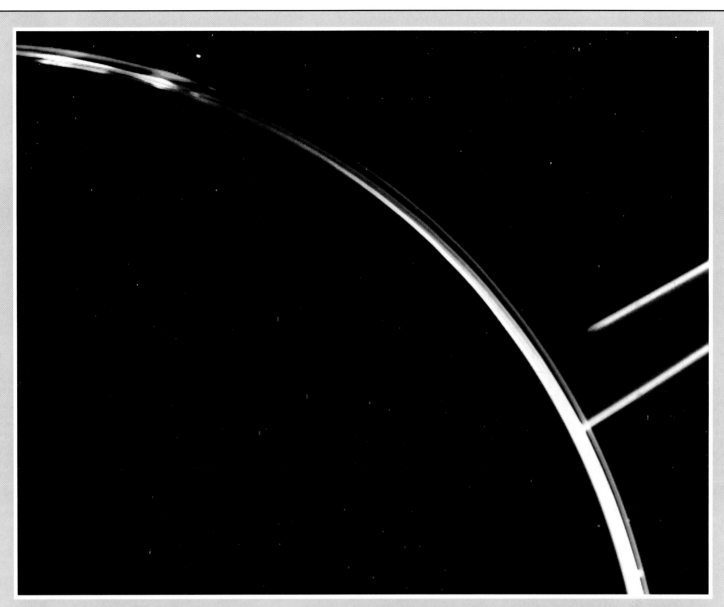

Left: The chemical composition of Jupiter's interior *(top)* as compared to Saturn's interior *(bottom)* based on Voyager data. *Above:* Jupiter's faint ring system. *Right:* A close-up of Jupiter's atmospheric disturbances.

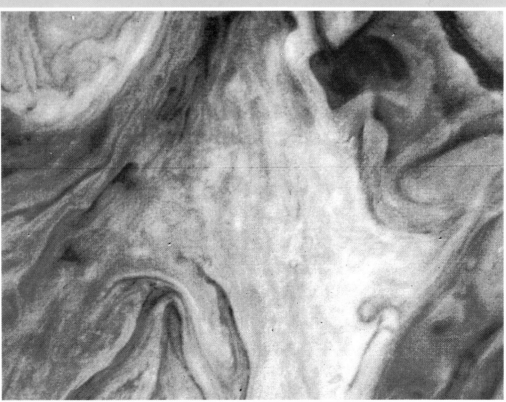

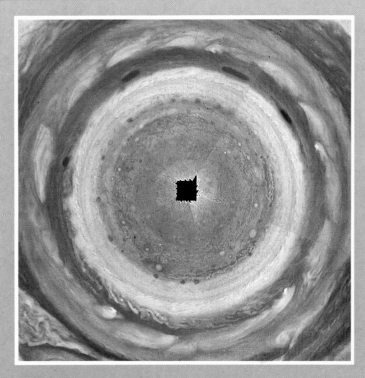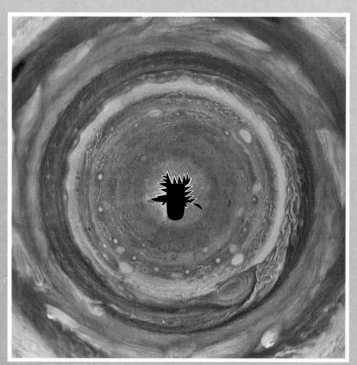

half a degree and they are all between 250,000 and 700,000 miles from Jupiter.

The third group was discovered in the twentieth century prior to the Voyager project, and they all have diameters of less than 105 miles. They all orbit in a plane whose orbital inclination is between 26 and 29 degrees, and they all are between 6.9 and 7.2 million miles from Jupiter. Like the third group, the final group were all discovered in the twentieth century prior to the Voyager project, and they all have diameters of less than 17 miles. They all exist in a plane whose orbital inclination is between 147 and 163 degrees and they are all between 12.8 and 14.7 million miles from Jupiter.

The inner group of Jovian moons is dominated by Amalthea, discovered in 1892 by **Edward Emerson Barnard (1857-1923)**. Amalthea is named for the goat-like nurse of Zeus, the Greek equivalent of the Roman god Jupiter. In Greek mythology, Zeus broke off one of Amalthea's horns and endowed it with the power to be filled with anything the owner wished, causing Amalthea to become associated with prosperity and riches. The innermost Jovian moon, Metis is named for the Greek god of prudence, wife of Zeus and mother of Athena. The other two inner Jovian moons, Adrastea and Thebe, were originally designated 1979 J1 and 1979 J2 and were the first two new moons in the Solar System to be discovered by the Voyager project. Adrastea is named for the mythical Greek king of Argos, whose daughter married Polynices of Thebes, who had been exiled by his brother. Adrastea (or Adrastus) led several primitive expeditions against Thebes.

The Galileans, Jupiter's second group of moons, were all discovered by Galileo (hence their name) during the first month (December 1609-January 1610) of the great astronomer's telescopic survey of the heavens. Aside from the Earth's Moon, they were the first planetary satellites to be observed and are easily seen from Earth with a telescope of moderate power. They are not only much, much larger than any other Jovian moons, they are among the largest in the Solar System, and Ganymede is *the* largest moon in the Solar System.

The next group of Jovian moons are Leda, Himalia, Lysithea and Elara. Himalia and Elara were discovered by **CD Perrine** at Lick Observatory between November 1904 and February 1905. Lysithea was discovered by SB Nicholson at Lick Observatory in 1938, while Leda, discovered by Charles Kowal at Mount Palomar in 1974, was the last Jovian moon to be discovered from Earth. Leda is named for the queen of Sparta, who was the mother (by Zeus in the form of a swan) of Helen, Castor and Pollux.

The outermost group of Jovian moons, Ananke, Carme, Pasiphae and Sinope, are the moons most distant from their mother planets of any known in the Solar System. The four moons of the Jovian outer group also all orbit in a *retrograde* motion. The only other moons anywhere in the Solar System that move in retrograde motion are Saturn's Phoebe and Neptune's Triton. **SB Nicholson** discovered Ananke and Carme with the reflector telescope at Mount Wilson, California in September 1951 and July 1938 respectively. Pasiphae and Sinope were discovered by **PJ Melotta** at Greenwich, England in January 1908 and July 1914 respectively. Ananke is named for a Greek cult goddess who shared a shrine at Corinth along with the goddess Bia. Horace later assigned her the Latin name Necessitas.

The innermost of Jupiter's Galilean moons, Io, is one of the most intriguing bodies in the Solar System. In 1979, photographs returned by the two Voyager spacecraft revealed a very dynamic world whose surprising characteristics were beyond anything that had previously been imagined. The most volcanically active body in the Solar System, Io is the only place besides Earth where volcanic eruptions have actually been observed.

Named for the maiden in Greek mythology who became a lover of Zeus only to be turned into a cow by Hera, the moon called Io is a unique world. Io's surface, with its brilliant reds and yellows that remind one of a giant celestial pizza, is a crust of solid sulfur 12 miles thick that floats on a sea of molten

(continued on page 131)

Opposite, clockwise from upper left: **The northern and southern hemispheres of Jupiter taken by Voyager 1 from directly above the poles, and the region of Jupiter extending from the equator to the southern polar latitudes near the Great Red Spot (GRS) taken by Voyager 2. The disturbed region to the west of the GRS has changed since the equivalent Voyager 1 image.** *Right:* **Two of Jupiter's four Galilean satellites are visible in this Voyager 1 image: Io (orange) and Europa (whitish).**

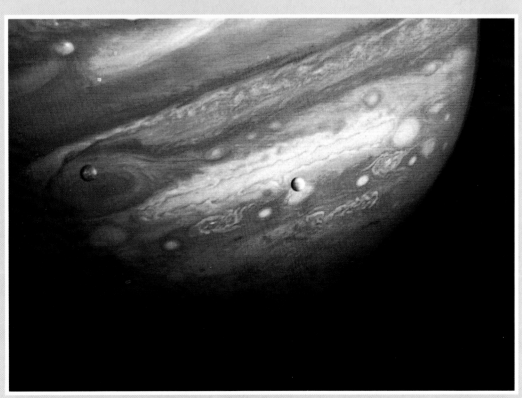

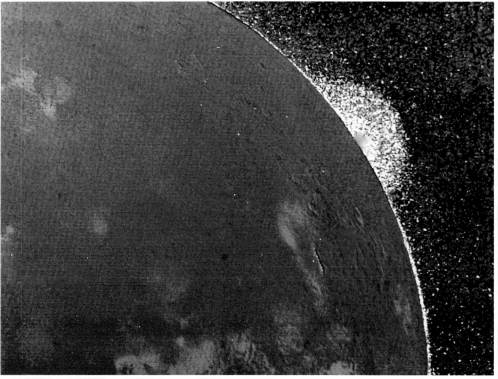

Above: Io as seen by Voyager 1 on 4 March 1979. The donut-shaped feature is an erupting volcano, a common occurance on Io. It is the first body in the Solar System outside Earth where active volcanism has been observed. *Left:* This color reconstruction of one of Io's erupting volcanos. The region that is brighter in ultraviolet light (blue in this image) is much more extensive than the denser, bright yellow region near the center of the eruption. *Right:* Europa is the smallest and the brightest of the four Galilean satellites. The complex pattern on the surface suggests that the icy surface is fractured and that the cracks are filled with dark material from below.

sulfur. This sea is, in turn, thought to cover silicate rock which is at least partially molten. The tidal effect of so massive a body as nearby Jupiter is thought to cause the crust to rise and fall on the molten sulfur by as much as sixty miles. It is through this heaving sulfur crust that Io's volcanos have burst.

The red and yellow sulfurous surface is marred by dozens of jet black volcanos whose violent eruptions surpass (both in magnitude and frequency) anything seen on Earth. During its brief encounter with Io in March 1979, Voyager 1 observed the eruptions of no fewer than eight volcanoes, with the plume above Pele reaching an altitude of 174 miles. When Voyager 2 turned its cameras toward Io four months later, it was able to observe eruptions still taking place at all the volcanos discovered by Voyager 1 except two.

In these violent eruptions, sulfurous material is belched from Io's liquid mantle at speeds of up to 3280 feet per second—many times the recorded velocity of the Earth's volcanos. The reason for the altitudes of the plumes and the velocity of the particles is due in part to the weaker gravity on Io—whose mass is less than two percent that of the Earth. The sulfur particles fall to the surface relatively fast, however, because there is no atmosphere on Io in which they could become suspended, and hence no winds to blow them in great billowing clouds of ash across the land, as was the case on Earth following the recent eruptions of Mount Etna and Mount St Helens.

Each of Io's volcanic eruptions dumps 10,000 tons of sulfur onto the moon's surface. In extrapolation, this would account for 100 billion tons of sulfur deposits per year. This is enough to cover the entire surface with a layer of sulfur 'ash' a foot thick in 30,750 years—a relatively short time, geologically speaking. Combined with surface flows, Io could very well be completely resurfaced with a foot-thick layer in as short a period of time as 3100 years, giving this pizza-colored moon the youngest solid surface in the Solar System aside from Earth, and there are parts of Earth that change *less* over time than does much of Io. In fact, there were many noticeable changes in Io's surface—particularly around the volcano Pele—in just the four months between March and July 1979. This 'ever-youthful' surface accounts for the complete absence of meteorite impact craters on Io.

Surrounding the black volcanic caldera are black fan-shaped features that are the result of liquid sulfur that cools rapidly as it reaches Io's frigid surface. South of Loki, the Voyager imaging team discovered a U-shaped molten sulfur lake 125 miles across that had partially crusted over. It was detected by its surface crust temperature of about 65 degrees Fahrenheit—compared to the surrounding surface temperature of less than -230 degrees Fahrenheit. This lake has certainly cooled and solidified by now, while other molten sulfur lakes have no doubt formed elsewhere.

(continued on page 132)

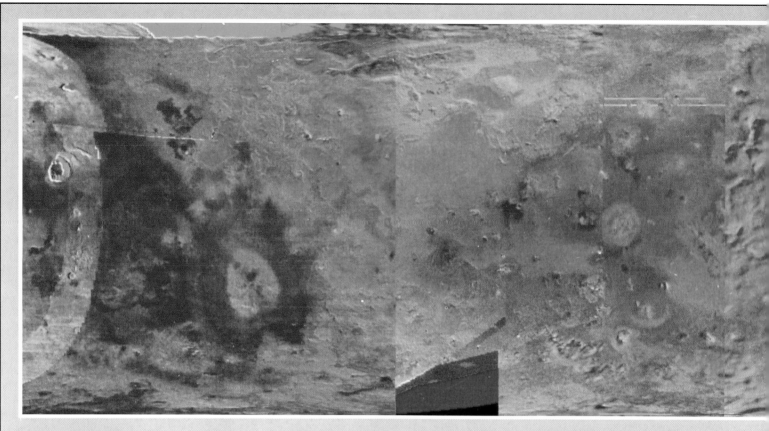

The moon Europa has a surface that is probably composed entirely of water ice. Because this surface is marred only by three definite impact craters, Europa is a mystery. The absence of impact craters on Io is explained by its violently active surface, but Europa has an extremely smooth and apparently inactive surface. Named for the Phoenician princess abducted to Crete by Zeus, Europa has a highly reflective surface that probably remained in a slushy semi-liquid state until relatively recently. This is at present the only explanation for its lack of meteorite craters.

The smooth surface is, however, not without features and these, too, present part of Europa's mystery. The features include long black linea, or lines, reminiscent of Percival Lowell's 'canals' on Mars. These features, which are up to 40 miles wide and stretch for thousands of miles across the surface, defy explanation. They appear to be cracks in the lighter surface, but they have no depth and thus they can only be described as 'marks' on the surface.

Another peculiar feature of Europa's icy surface is the presence of numerous dark macula, or spots, distributed across the surface. While the three impact craters range from 11 to 15.5 miles across, the macula are generally smaller than six miles.

The most unusual features on Europa are the flexus. Light colored, scalloped ridges, the flexus are much narrower and somewhat shorter than the linea. They are, however, more regular in width and in the regularity of their scallops or cusps.

Europa is thought to have a relatively large silicate core with a layer of molten silicate above that, which is, in turn, covered by a layer of liquid water perhaps 60 miles deep. Above this is the strange, icy crust, which is roughly 40 miles thick. Ganymede is the largest moon in the Jovian system and is, indeed, the largest moon in the entire Solar System. Ganymede, like Callisto, is composed of silicate rock and water ice, and thus these bodies came to be dubbed 'dirty snowballs.' Named for the cup-bearer of the Greek gods, Ganymede has an ice crust

Above: This false-color photo mosaic of Io shows subtle variations in the surface deposits. *Below:* **Jupiter with four moons in order of their approximate distances from Jupiter: Io, Europa, Ganymede and Callisto.** *Opposite above:* **The dark feature on Ganymede known as Galileo Regio, is marked by curved ridges.** *Opposite below:* **Valhalla, an enormous impact basin in the northern hemisphere, is Callisto's largest feature.**

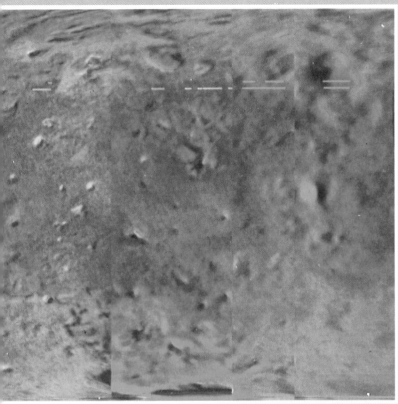

that is roughly 60 miles thick. This crust, in turn, floats upon a mantle of slushy, partially-liquid water that is roughly 400 miles deep. Beneath Ganymede's mantle is a heavy silicate core.

Like Earth's rocky crust, Ganymede's icy crust is divided into plates which shift and move independently, interacting with one another along fracture zones, resulting in geologic activity that is very much like what has been observed on Earth. Mountain ranges 10 miles across and 3000 feet high have formed on Ganymede as a result of the pressure of ice plates against one another. The surface of Ganymede is characterized by mountainous terrain and ancient dark plains, the largest of which is the region known as Galileo Regio. The dark plains are, in turn, marked by a wrinkled or grooved terrain consisting of a semicircular system of parallel, curved ridges six miles wide, 325 feet high and approximately 40 miles apart. These grooves are the remnants of an ancient impact basin that has long since been obscured by subsequent geologic activity.

A great number of smaller impact craters have been identified on Ganymede's surface, with many of them showing white 'halos.' These are evidence of water having been splashed up through the crust after each fiery meteoroid smashed its way through the surface.

Because of the size of Ganymede and the presence of liquid water, it once was suggested that there might be a tenuous atmosphere of water vapor and free oxygen (with the latter being formed by the effect of sunlight on the former). No evidence of an atmosphere was detected by the Voyager spacecraft in 1979, and indeed, if one were present, its atmospheric pressure less than one hundred billionth of Earth's.

Callisto has a water ice and silicate rock composition like that of Ganymede, but unlike Ganymede, Callisto's ice and silicate soil surface shows no sign of any geologic activity. The only surface feature on this 'dirty snowball' is a mass of hundreds upon hundreds of impact craters. The largest of these is Valhalla, a huge impact basin in Callisto's northern hemisphere that measures 1860 miles across. The probable reason for the lack of geologic activity is that Callisto's icy crust is more than 150 miles thick, and thus is not prone to break into plates as Ganymede's has. It is also much farther from the tidal effects of Jupiter's gravity. Beneath the solid ice crust is a slushy mantle 600 miles in depth, and beneath the mantle is a heavy silicate core. Thus Callisto is thought to be identical to Ganymede in terms of its composition, but with a thicker, and hence more geologically inert, crust.

Callisto was named for a Greek nymph favored by Zeus and turned into a bear by the jealous Hera. With temperatures ranging between -200 and -300 degrees Fahrenheit—roughly the same temperature as Hera's heart—there is little likelihood of an atmosphere hanging over Callisto's frigid wastes.

A PORTRAIT OF SATURN

In terms of sheer size, Saturn dwarfs all the other planets except Jupiter, but its incredible system of rings puts it visually in a class by itself. Named for Jupiter's father, the original patriarch of Roman gods, Saturn is the outermost of the planets visible from Earth with the unaided eye.

Saturn's composition is very much like that of Jupiter. There is probably a solid core composed of iron and silicates that measures about 9300 miles in diameter, which is covered by a layer of water kept in solid ice form by the pressure of successive higher layers of metallic hydrogen and liquid molecular hydrogen. Like Jupiter, nearly 80 percent of Saturn's mass is hydrogen, the simplest element, with most of the remainder taken up by helium, the second simplest and second most common element in the solar system, so Saturn may have had the same star-like ancestry as Jupiter. Major components of the core are iron (0.2 percent) and silicates (0.1 percent), as well as oxygen (1 percent), which also is present, with hydrogen, in the water ice. The remaining elements are the inert gas neon (0.2 percent) which, along with organic gases composed of nitrogen (0.1 percent) and carbon (0.4 percent), comprise Saturn's atmosphere.

Saturn's atmosphere contains many of the same gases that are present in the atmospheres of Jupiter, Uranus and Neptune. The most common are methane and ammonia, but there are also traces of phosphene and more complex organic compounds such as propane, ethane, acetylene and methylacetylene. Saturn's atmosphere is a good deal smoother, hazier and less choppy than Jupiter's, with relatively few distinct features to parallel Jupiter's brown, white and 'Great Red' spots.

Saturn's cloud tops are smoother than Jupiter's, probably due to weaker gravity (because of smaller mass) and lower temperatures. At these colder temperatures the condensation point of the chemicals in the atmosphere would be reached in regions of higher pressure and, hence, at lower altitudes within the atmosphere. However, Saturn's atmosphere is probably characterized by the same visible turbulence as Jupiter's, but at lower altitudes below a layer of ammonia haze. Saturn's atmosphere is characterized by horizontal bands alternating between those with westerly—or the rarer easterly—prevailing winds. These winds have speeds of up to 900 mph, with the greatest westerly velocities being recorded within five degrees latitude of the equator. The largest feature in Saturn's atmosphere is **Anne's Spot**, a pale red feature in the southern hemisphere that is similar to, though smaller than, Jupiter's Great Red Spot. Like the latter, Anne's Spot is thought to be composed of phosphene that is brought high into the upper atmosphere by spiraling convection currents.

Saturn's ring system, while not absolutely unique, is certainly the planet's outstanding feature. Galileo first observed the rings in 1610, but Saturn happened to be oriented so that the great Italian astronomer was viewing them *nearly* edge-on, and thus it wasn't clear what they were.

Galileo at first thought he had discovered two identical moons of the scale that he had found near Jupiter. However, these 'moons' did not rotate or change position and Galileo was mystified. He wrote to the Grand Duke of Tuscany that 'Saturn is not alone but is composed of three, which almost touch one another and never move nor change with respect to one another. They are arranged in a line parallel to the

Left: A true-color image of Saturn and three of its moons. *Right*: The Saturnian ring system completely baffled early astronomers. It wasn't until the Voyager 1 flyby that astronomers discovered the extent of Saturn's rings.

zodiac, and the middle one [*Saturn itself*] is about three times the size of the lateral ones' [*actually the outer edges of the rings*].

Two years later the plane of the rings was oriented *directly* at the Earth and the 'lateral moons' seemed to disappear entirely. Galileo was completely baffled, but no less so than when they reappeared in 1613. In December 1612, Galileo had written 'I do not know what to say in a case so surprising, so unlooked for and so novel. The shortness of the time, the unexpected nature of the event, the weakness of my understanding, and the fear of being mistaken have greatly confounded me.'

More than a decade after Galileo's death in 1655, the Dutch astronomer **Christiaan Huygens (1629-1695)** solved the riddle. Using a telescope more powerful than that which was available to his predecessor, Huygens figured out that the mysterious objects were rings around Saturn and the reason for their 'disappearance' in 1612. He also went on to calculate that the rings would be oriented in this way on a 150-month cycle, and that at opposing ends of the same cycle almost the entire ring would be visible from Earth. It has since been determined that the cycle actually alternates between periods of 189 and 165 months. Huygens also discovered Saturn's largest moon, Titan.

In 1671, the Italian-born and naturalized French astronomer **Giovanni Domenico Cassini (1625-1712)** began his own observations of the ringed planet. Cassini discovered a second moon of Saturn, Iapetus, in 1671, and in 1675 he determined that the 'ring' around Saturn was not a single band, but a pair of concentric rings. These two rings would come to be known as the A Ring and B Ring, with the space between them
(continued on page 136)

appropriately named the Cassini Division. In 1837, **Johann Franz Encke (1791-1865)** at the Berlin Observatory tentatively identified a faint division in the A Ring. This division was confirmed in 1888 by **James Keeler (1857-1900)** of the Allegheny Observatory in the United States. Subsequently, this division has been known as either the 'Keeler Gap' or (more often) as the 'Encke Division.'

The first manmade spacecraft to venture close to Saturn was the American Pioneer 11 in September 1979. Prior to this time, there were only three known rings of Saturn, each lettered in the order of their discovery from A through C. Pioneer 11 helped Earth-based astronomers identify a fourth ring, which is now known as the F Ring. When the American Voyager 1 Spacecraft first approached Saturn in November 1980, the spectacular photographs that were beamed back to Earth revealed that there were not just four, six, or even a dozen rings in Saturn's ring system; rather, there were literally thousands of rings, with each known ring itself composed of hundreds or thousands of rings, with faint rings identified even within the Cassini Division.

At a point 19,600 miles above Saturn's cloud tops, the C Ring merges into the B Ring without a major gap. The B Ring, which is 15,800 miles wide, is the brightest of Saturn's main rings and also the widest, except for the virtually invisible E Ring. The 2800 mile-wide Cassini Division separates the B Ring from the 9000 mile-wide A Ring, the second brightest of Saturn's rings.

The bright, yet tenuous F Ring is located just 2300 miles beyond the outer edge of the A Ring, and the narrow G Ring is located 20,700 miles from the A Ring. The E Ring is a very, very faint 55,800 mile-wide mass of particles that begins 91,000 miles from Saturn's cloud tops and extends past the orbit of the moon Enceladus.

Saturn's rings are composed of silica rock, iron oxide and ice particles, which range from the size of a speck of dust to the size of a small automobile. They range in density from the nearly opaque B Ring to the very sparse E Ring. Theories about the Ring system's origin generally fall into two camps. One theory, originating with Edward Roche in the nineteenth century, holds that the rings were once part of a large moon whose orbit decayed until it came so close to Saturn as to be pulled apart by the planet's tidal or gravitational force. An alternate to this theory suggests that a primordial moon disintegrated as a result of being struck by a large comet or meteorite. The opposing theory is that the rings were never part of a larger body, but rather they are nebular material left over from Saturn's formation 4.6 billion years ago. In other words, they were part of the same pool of material out of which Saturn formed, but they remained separate and gradually formed into rings.

Not only do its spectacular rings set Saturn apart from other planets, but so too does its complex system of more than 20 moons. Saturn's moons range in size from huge Titan, once thought to be the Solar System's largest moon, to the family of tiny moons that were discovered in photographs taken by the Voyager Spacecraft in 1980. Though the moons of Saturn are no less diverse in character than those of Jupiter, they are generally smaller and, with the exception of the two outermost (Iapetus and Phoebe), their orbital inclination is within 1.5 degrees of that of the rings. With the exception of Phoebe, the moons are synchronous, like Earth's moon, meaning that the same side faces Saturn at all times. The western hemi-

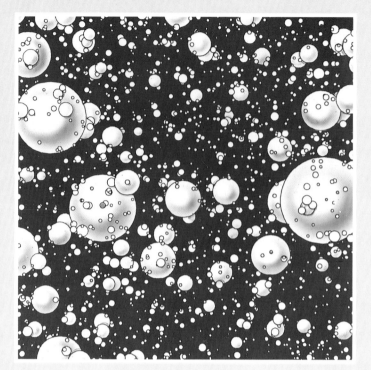

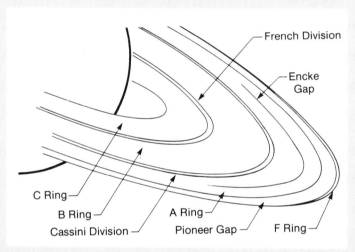

At top: **Particles in a cross-section of the A Ring range in size from marbles to beach balls.** *Above:* **A schematic of the ring system.** *Opposite:* **This color image suggests differences in the chemical composition of the rings.**

spheres, which face in the direction of their orbital paths, are called leading hemispheres, while the eastern are called trailing hemispheres.

Having discovered Neptune's moon Triton in 1846 with his reflector telescope with a two-foot aperture, **William Lassell (1799-1880)** discovered Saturn's moon Hyperion in 1848, which was also independently found by **William Cranch Bond (1789-1859)** with the 15-inch refractor telescope at Harvard College Observatory.

In 1977, when the two Voyager Spacecraft were launched from Earth, the ringed planet was known to have nine moons. Five of these were discovered prior to 1700 (with four found by Giovanni Cassini, the man who discovered that Saturn had multiple rings), and only two were discovered after 1800. By the time that the Voyager data was digested in 1982, Saturn was known to have 17 moons and four to six additional Lagrangian co-orbital satellites. A co-orbital is one of a group of moons that share a single orbital path, while Lagrangian satellites (named after the eighteenth century astronomer

whose mathematical theory postulated their existence) are small co-orbitals that exist in the orbit of a larger moon and remain 60 degrees ahead or 60 degrees behind it in the orbital path. Saturn's most recently discovered moons are so small and so close to the planet that it is almost hard to know where to draw the line between moons and ring particles. This also makes it harder to find such bodies visually against the brilliant rings.

The innermost of Saturn's moons is Atlas, which is named for one of the Titans of Greek mythology who was condemned to support the weight of the universe on his shoulders. Atlas is also informally known as the A Ring Shepherd Moon because of its role in shepherding the nearby outer A Ring particles and, in a sense, *defining* the outer edge of the A Ring. The next two moons, Prometheus and Pandora, are named respectively for the Titan of Greek mythology who stole fire from Olympus to give it to man, and for the woman who was bestowed upon man as a punishment for Prometheus having stolen fire. Prometheus and Pandora are known informally as F Ring Shepherd Moons because of their positions on either side of the F Ring and their role in defining that ring. The two moons may also be responsible for the kinks and braiding observed in the F Ring.

Well beyond the F Ring, but inside of the G Ring, are Epimetheus and Janus, which are the first of the several groups of co-orbital moons and the only group of co-orbitals not to be of the Lagrangian type. The centers of these two bodies, and hence the 'center lines' of their orbital paths, are offset by only 30 miles, a distance narrower than the radius of either! Epimetheus is named for the brother, in Greek mythology, of Prometheus, who accepted Pandora (a gift from Zeus) as his wife, despite the warning of his brother. As Prometheus had warned, Pandora opened the infamous box, releasing all the evils within. Janus, the two-faced Roman god of doorways, became the namesake of a moon that was thought to have been identified in 1966 at a distance of 105,000 miles from Saturn. The existence of this 'first' Janus was disproven, but the name was reassigned to 1980 S26, which was discovered in a nearby orbit.

In July 1990, a new moon orbiting Saturn was discovered by **Dr Mark Showlater**, a researcher at NASA's Ames Research Center at Mountain View, California. Showlater found the small, bright object while analyzing images taken by Voyager 2. He had used a comupter program he had written to sort

(continued on page 138)

thorough 30,000 images sent back to Earth during the Voyager/Saturn encounters in 1980 and 1981.

The discovery brought to 18 the number of known Saturnian moons. One new moon was found to have a diameter of only 12 miles and was temporarily designated 1981S13. It was documented as orbiting in Encke's Division, in Saturn's outermost major ring, the A Ring.

Now Saturn's smallest known satellite, the moon pushes material away from its orbit and is believed to cause the 200 mile-wide Encke's Division. 'The same shepherding effect,' Showlater said, 'produced Saturn's thin, outermost F Ring, with moons orbiting on either side of the ring, constraining the material.'

A wavy pattern in the ring material on both sides of Encke's Division was first noticed by Ames' ring expert **Dr Jeff Cuzzi** while studying basic ring structure in the mid-1980s. Adapting a theory from galactic cynamics, he and **Dr Jeffrey Scargle**, also at Ames, suggested that the disturbance was caused by an unseen asteroid-sized moon in the gap.

Discovered in 1789 by the German-born, but naturalized English, astronomer **William Herschel (1738-1822)**, Mimas is scarred by a huge impact crater that bears his name. The walls of the crater Herschel average 16,000 feet, and a huge mountain at the crater's center rises nearly 20,000 feet from its crater floor. This huge crater is centered precisely on the equator and has a diameter one third the diameter of Mimas itself. Herschel is truly the standout feature on Mimas, as none of the other impact craters observed on the surface have anywhere near half its diameter. Mimas's surface is also characterized by valleys, or chasma, which tend to run in a parallel pattern from southwest to northwest. These uniform valleys are generally 60 miles long, one mile deep and six miles wide. They are thought to be fracture zones which date from the impact which formed Herschel. Mimas is thought to be composed of 60 percent water ice with roughly 40 percent silicate rock, making it of the 'dirty snowball' class of moons found in much of the outer solar system. In 1982, **Stephen Synnott** at the Jet Propulsion Laboratory in Pasadena, California, identified a probable co-orbital moon within the orbit of Mimas.

Enceladus is named for the giant who rebelled against the gods of Greek mythology, and who was subsequently struck down and buried on Mount Etna. Discovered by William Herschel in 1789 at the same time that he identified Mimas, Enceladus is the most geologically active of Saturn's moons.

Like Mimas, Enceladus is composed mostly of water, with the remaining roughly 40 percent of material being silicate rock. Enceladus, however, has a much more complex surface which is divided between vast and ancient fields of impact craters, large smooth plains and complex mountain ranges. The latter are typical of the type of fracture zones that characterize the surfaces of Jupiter's moon Ganymede or Uranus' moon Miranda. These features, including ridges and valleys, were possibly formed by the same sort of pressure between separate surface plates that is responsible for the silicate rock mountain ranges on Earth and the ice mountain ranges on Ganymede. The smooth plains on Enceladus are further evidence of fracture zones because they were possibly formed by liquid water welling up from the interior and spilling out through fissures and faults, forming lakes in lowland areas. These lakes covered older fields of impact craters, and when

(continued on page 140)

Clockwise from above: *Strange cloud formations in Saturn's atmosphere; Saturn, Dione, Tethys, Mimas, Enceladus, Rhea and Titan; Mimas; the 'spoke' feature in the B Ring; striations in Saturn's clouds are caused by latitudinally oriented winds, which can reach velocities of 900 mph.*

THE VOYAGER PROJECT 139

the water froze the lakes became the relatively smoother and much newer plains which we see today. The most notable feature on Tethys is the mysterious Ithaca Chasma, an enormous rift canyon that runs from near the north pole all the way to the south pole. With an average width of 60 miles and an average depth of three miles, Ithaca Chasma dwarfs the Earth's Grand Canyon in both scale and absolute terms. In the scale of the Earth, the equivalent of Ithaca Chasma would be like having a 40-mile deep trench as wide as the state of Colorado, extending from Nome, Alaska, to the southern tip of Argentina. Of uncertain origin, this huge canyon was probably crated when Tethys cooled after it was first formed.

Discovered in 1684 by **Giovanni Domenico Cassini (1625-1712)**, Tethys is named for the Greek sea goddess who was both the wife and sister of Oceanus. Like the smaller Mimas and Enceladus, Tethys is a 'dirty snowball' composed mostly of water ice, with silicate rock as a secondary component. Most of its surface is marred by impact craters, but they appear somewhat softened—as though the surface had warmed slightly at some point since the formation of the craters and that partial melting had taken place.

Dione's surface is darker than any of Saturn's other ice/rock moons, indicating that there are large regions of exposed rock. Dione's surface is characterized by impact craters common to both icy and rocky surface areas. The craters are generally smaller than 25 miles across, but Amata, the largest, measures nearly 150 miles in diameter. Amata is coincidentally located at the center of an eastern hemisphere pattern of unusual light colored streaks. While the streaks on Enceladus and Tethys are very sharply defined, as though cut with a knife, Dione's streaks are wispy—as though they were painted with an air brush. These streaks are probably cracks or fissures through which liquid water seeped and refroze over time. Amata appears to have been partially inundated by this seepage, and it has been suggested that whatever impact created Amata may have also played a role in the formation of the wispy streaks, because many of them seem to radiate from the crater. Discovered by Giovanni Cassini in 1684, Dione is named for the mother of the Greek goddess Aphrodite. Like the moons closer to Saturn, Dione is composed of silicate rock and water ice. However, while the inner moons have a predominance of water ice, Dione is at least half rock.

The second largest of Saturn's moons, Rhea was discovered by Cassini in 1672 and was named for the wife of Kronos, who according to Greek mythology, ruled the universe until dethroned by his son Zeus. In Roman mythology, as well as in astronomical nomenclature, Rhea is identified with Saturn because Saturn is the father of Jupiter, the Roman equivalent of Zeus.

Rhea the moon is composed of an equal mixture of water ice and silicate rock, with a possible core of denser material, or perhaps solid rock. The moon's leading hemisphere is solid water ice with intermittent patches of frost, while the darker, trailing eastern hemisphere shows signs of the same wispy surface detail that was observed more minutely by the Voy-
(continued on page 142)

Left: Enceladus as seen by Voyager 2 from a distance of 74,000 miles. The uncratered surface of this moon of Saturn leads astronomers to believe that it has recently, in geological time, experienced internal melting. *Above:* Iapetus, the outermost of Saturn's satellites. *At top:* Rhea is the second largest of Saturn's moons. *Right above:* Dione, discovered in 1684, is at least half rock. *Right:* Tethys is among the 'dirty snowball' moons that are commonly found in the outer Solar System.

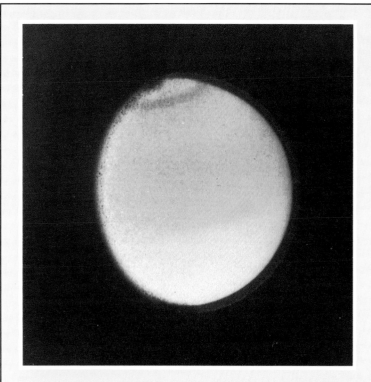

ager spacecraft on Dione. The icy western hemisphere is characterized exclusively by thousands of impact craters, while there seem to be fewer craters in the eastern hemisphere and in the equatorial regions of both hemispheres. The largest crater, Izanagi, located deep in the southern hemisphere near the prime meridian, has a diameter of nearly 140 miles.

Larger than either of the planets Mercury and Pluto, Titan is the second largest moon in the Solar System (after Ganymede). Once thought to be the largest, it was named for the family of pre-Olympian Greek gods whose name implies colossal size. Titan was discovered by the Dutch astronomer **Christiaan Huygens** in 1655, making it one of the first moons to be discovered in the Solar System. Though no longer on the throne as the Solar System's largest moon, it certainly must rank as one of the most spectacular.

Titan is the only known moon with a fully developed atmosphere that consists of more than simply trace gases. It has, in fact, a denser atmosphere and cloud cover than either Earth or Mars. This cloud cover, nearly as opaque as that which shrouds Venus, has prevented the sort of surface mapping that has been possible with the other major moons of the outer Solar System, but its presence has only served to make

Left: Titan was one of the first moons to be discovered in the Solar System. It is the largest of Saturn's satellites, and once was thought to be the largest in the Solar System. *Left below:* Titan is the only moon in the Solar System to have developed a significant atmosphere. In this false-color image of Titan, the layers of methane and nitrogen haze occur at altitudes of 124, 233 and 310 miles above the limb. *Right:* Artist Don Davis painted this rendition of Voyager's closest approach based upon a comuputer-assembled simulation of the spacecraft's journey through the Saturn System. The Sun is seen in the upper left sky. The main group of rings from the innermost D Ring to the A Ring reveal the hundreds of ringlets discovered as well as the 'braided' F Ring and the dusty G Ring.

Titan all that much more intriguing. (Early astronomers mistook this dense atmosphere for Titan's actual surface, and it was through this mistake that Titan was once considered to be the Solar System's largest moon.)

Titan's atmosphere is extremely rich in nitrogen, the same element that makes up the greatest part of the Earth's atmosphere. Other major components of Titan's atmosphere are hydrocarbon gases, such as acetylene, ethane and propane, with methane being the most common of the hydrocarbons. While these gases are also to be found in Saturn's own atmosphere, Titan's atmosphere contains four times the ppm concentration of ethane and 150 times the concentration of acetylene. Titan's atmosphere probably includes broken methane clouds at an altitude of about 25 miles, with the dense, smoggy hydrocarbon haze stretching up to an altitude of nearly 200 miles, where ultraviolet radiation from the Sun converts methane to acetylene or ethane.

Why Titan was able to develop an atmosphere, while Ganymede and Callisto (bodies of similar size) did not, is a matter of conjecture. It has been theorized that all the Solar System's largest moons had similar chemistry in the beginning, but Titan evolved in a colder part of the Solar System—farther away from the Sun and from Jupiter—when it almost became a primordial star. Thus the hydrocarbon gases were able to exist as solids on Titan, while the gases of Jupiter's Galilean moons dissipated into space, leaving only water and rock. Saturn's other moons meanwhile, were never large enough to have sufficient gravity to hold an atmosphere.

The presence of nitrogen, a hydrocarbon atmosphere and water indicate that Titan's surface is very much like that of the Earth four billion years ago before life evolved on the latter body. It has been suggested that this similarity to the prebiotic 'soup' that covered the Earth in those bygone days could presage a similar chain of events on Titan. The view from Titan's surface is one of an exciting, but inhospitable, world. Covered by the opaque haze, the sky would appear like a smoggy sunset on Earth or like a view from the surface of Venus. The atmospheric pressure on Titan's surface, while 1.6 times that of Earth is a good deal less than that of Venus. Titan's surface temperature of nearly 300 degrees Fahrenheit would permit methane to exist not only as a gas, but also as a liquid or a solid, in much the same way that water does on Earth. A picture is thus painted of a cold, orange-tinted land where methane rain or snow falls from the methane clouds and where methane rivers may flow into methane oceans dotted with methane icebergs. There is evidence of a 30 Earth-year seasonal cycle which *may* have permitted the development of methane ice caps that expand and recede like the water ice caps on Earth (and the water/carbon dioxide ice caps on Mars). Water ice is also present on Titan, beneath the methane surface features, and possibly extends up into the atmosphere in the form of ice mountains. Titan's mantle is, in turn, largely composed of water ice that gives way to a rocky core perhaps 600 miles beneath the surface. The absence of a magnetic field indicates that Titan has no significant amount of ferrous metallic minerals in its core.

A PORTRAIT OF URANUS

Uranus is named for the earliest supreme god of Greek mythology. The personification of the sky, mythical Uranus was both son and consort to the goddess Gaea and father of all the Cyclopes and Titans. The first outer Solar System planet to be correctly identified as such in historical times, Uranus was identified in 1781 by the German-born English astronomer **William Herschel** while he was working at Bath.

Uranus has an axial inclination of 98 degrees, a phenomenon that is unique in the Solar System. With such an axial inclination, Uranus is seen as rotating 'on its side' at a near right angle to the inclination of the Earth or Sun. The poles of Uranus, rather than its equatorial regions, are pointing alternately at the Sun.

Like Jupiter, Saturn and Neptune, Uranus is a gaseous planet with a distinct blue-green appearance, probably due to a concentration of methane in its upper atmosphere. In terms of size, it is smaller than Jupiter and Saturn, while being very close to the size of Neptune. Its solid core is composed of metals and silicate rock with a diameter of roughly 270,000 miles. Its core is, in turn, covered by an icy mantle of methane ammonia and water ice 6000 miles deep.

As with the other gaseous planets, the predominant elements in the Uranian atmosphere are hydrogen and helium, although the Voyager 2 observations in 1986 indicated that the atmosphere was only 15 percent helium, versus 40 percent, as originally postulated. Other atmospheric constituents

(continued on page 147)

Left: This montage of Voyager 2 images offers a view that one might see from a spacecraft above one of the huge canyons on Miranda's surface looking toward the blue-green Uranus shown here with an artist's conception of the planet's dark rings. *Right:* These two pictures of Uranus—one in true color (blue) and the other in false color—were compiled from the Voyager 2 spacecraft images taken seven days before its closest approach.

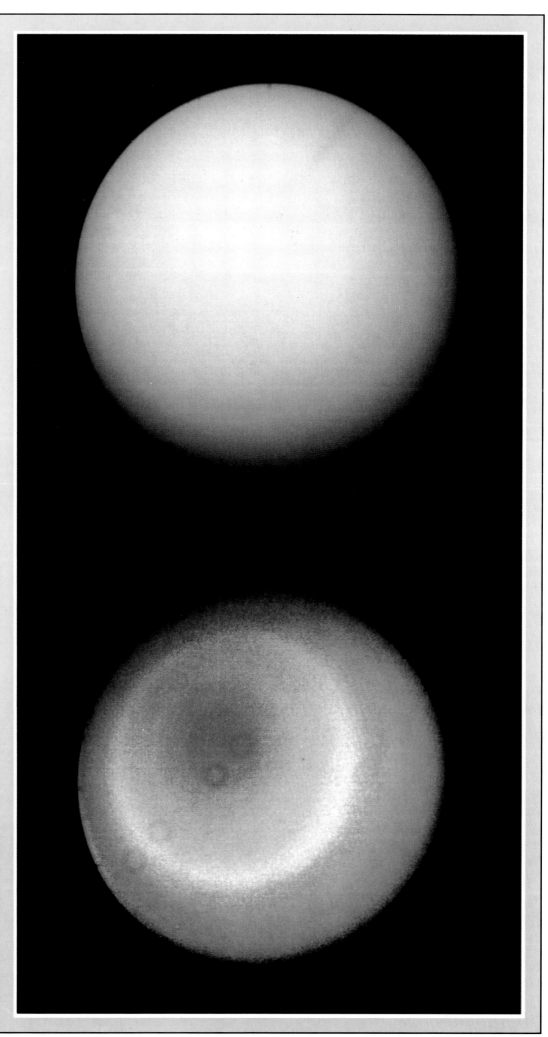

Left: This false-color view of all nine of Uranus' known rings. The brightest, or epsilon, ring at the top is neutral in color, and the other eight show color differences. Moving inward toward Uranus *(the bottom)* are the delta, gamma, and eta rings in shades of green and blue; the beta and alpha rings in lighter tones; and then a final set of three rings, known simply as 4, 5 and 6 in faint mauve tones. The fainter, pastel lines between the rings are products of the computer-enhancement process.

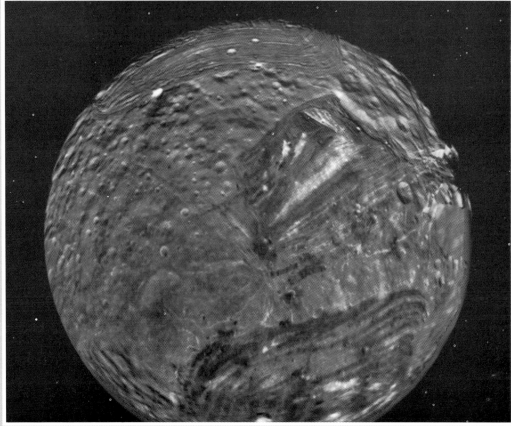

Above: Uranus *(center)* and its five major satellites from largest to smallest: Ariel, Miranda, Titania, Oberon and Umbriel. *Left:* Miranda, the innermost and smallest of the five major Uranian satellites. The terrain is both old—heavily cratered and rolling—and young—complex terrain characterized by sets of bright and dark bands, scarps and ridges. *Right:* Two slices of Uranus' epsilon ring assembled from data gathered by the Voyager photopolarimeter which measured the light from the star Sigma Sagittarii through the rings. The disparity in their sizes illustrates the accordian-like nature of the ring, which seems to contract and expand as it orbits the planet. The widths of these slices are 20 miles and 14 miles.

include methane, acetylene and other hydrocarbons. The clouds that form in this atmosphere are moved by prevailing winds that blow in the same direction as the planet rotates, just as they do on Jupiter, Saturn and Earth. The lowest temperature (-366 degrees Fahrenheit) is recorded at the boundary between troposphere and stratosphere. The coldest latitudes seem to be those between 15 and 40 degrees, but surprisingly, both the poles show similar temperatures whether or not they are sunlit!

Uranus, like Jupiter and Saturn, has a system of rings, of which the first nine were discovered by Earth-based observers in 1977. In 1986, Voyager 2 observed these in detail and identified two more. This ring system is much more complex than that of Jupiter, but less so than Saturn's spectacular system. The system around Uranus seems to be relatively young and probably did not form at the same time as the planet. The particles that make up the rings may be the remnants of a moon that was broken by a high velocity impact or torn apart by gravitational effects of Uranus. The widest ring known before Voyager 2 was the outermost ring, Epsilon—an irregular ring measuring 14 to 60 miles across. The outer edge of the system, the outer edge of the Epsilon ring, is sharply defined and is located 15,800 miles from the Uranus cloud tops. At this point the Epsilon Ring is just 500 feet thick, and surprisingly devoid of fragments with diameters below one foot.

Prior to the observations by Voyager 2, Uranus was known to have just five moons. Photographs returned by the spacecraft increased the number of known moons to 15, with all 10 of the newly-discovered moons located within the orbital paths of the original five. One of the new moons, Puck, was discovered by Voyager's cameras in late 1985, and the rest were discovered in the photos taken during the January 1986 Voyager flyby of the Uranian system. With the exception of Puck and Cordelia—the largest and smallest of the 'Voyager' moons—all of the newly discovered members of the group are very uniform in size, with diameters ranging between 31 and 37 miles.

The innermost of the moons are Cordelia, located between the Delta Ring and the Epsilon Ring, and Ophelia on the opposite side of the Epsilon. Thus straddling the Epsilon Ring, these two small bodies probably act like the shepherd moons of Saturn, controlling and defining the position and shape of the ring.

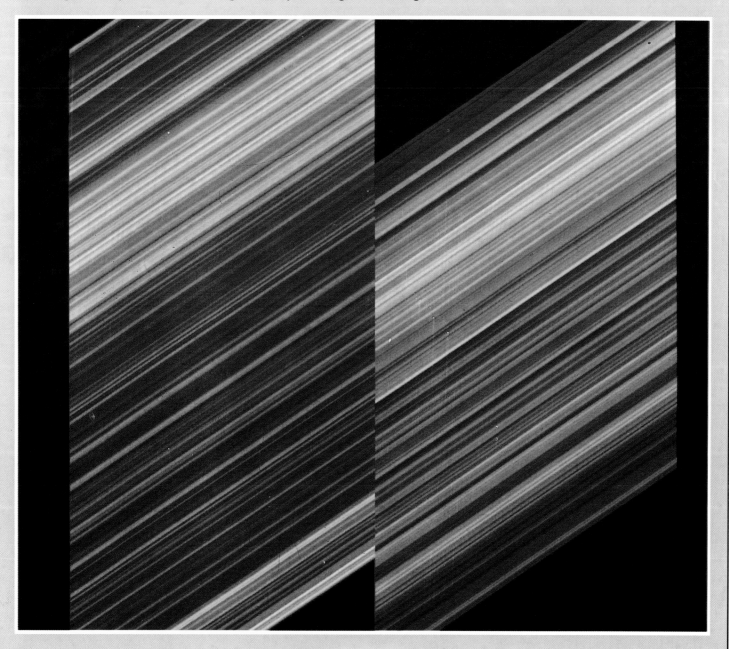

A PORTRAIT OF NEPTUNE

Located 2.8 billion miles from the Sun, Neptune is three times more distant from the Sun than Saturn and more than half again farther out as Uranus. At this distance, it takes 165 years for Neptune to complete one revolution around the Sun. Like Jupiter, Saturn and Uranus, Neptune is a giant gaseous orb. With a diameter of 30,642 miles, Neptune is a near twin of Uranus and is as close in size to this neighbor as Venus is to the Earth. Like Uranus, Neptune has a longer rotational period—16.3 hours—than Jupiter or Saturn, yet this rotational period is less than that of the terrestrial planets. Neptune also corresponds to Uranus in terms of its physical composition, which consists primarily of hydrogen and helium, with a methane and ammonia atmosphere.

After the discovery of Uranus, it took but 65 years before the existence of an eighth planet was confirmed. Galileo had sighted this object as early as 1613, but it was Johann Galle and Heinrich Ludwig D'Arrest who finally identified it as a planet in 1846. They named it Neptune, after the Roman sea god, because of its pale, sea-green color.

Prior to 1989, very little else was known about Neptune. The Voyager 2 spacecraft, which conducted close-up flybys of Jupiter, Saturn and Uranus in 1979, 1981 and 1986, turned its cameras on Neptune in June 1989 and flew to within 3044 miles of the planet on 25 August.

Located at a mean latitude of 22 degrees south and with an overall length of 30 degrees longitude, the **Great Dark Spot**

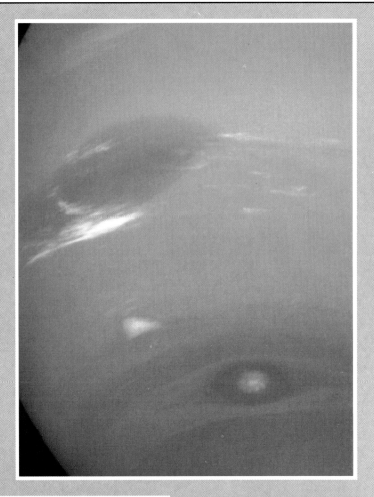

Above: Three features of Neptune, at the north (top) the Great Dark Spot (GDS), accompanied by bright, white clouds. To the south of the GDS is 'Scooter,' and further south is the feature called 'Dark Spot 2' or the Lesser Dark Spot, which has a bright core. Each moves at a different velocity, so it is only occasionally that they appear close together, such as at the time of this photo. *Left:* This Voyager picture provides obvious evidence of vertical relief in Neptune's bright cloud streaks. The linear cloud forms are stretched approximately along the lines of constant latitude and the Sun is toward the lower left. *Right top:* In this false-color image of Neptune, the pinkish clouds are at high altitudes while the dark blue areas are dark, deep-lying clouds, where visible light, but not ultraviolet light may penetrate. *Right bottom:* Triton, Neptune's largest satellite.

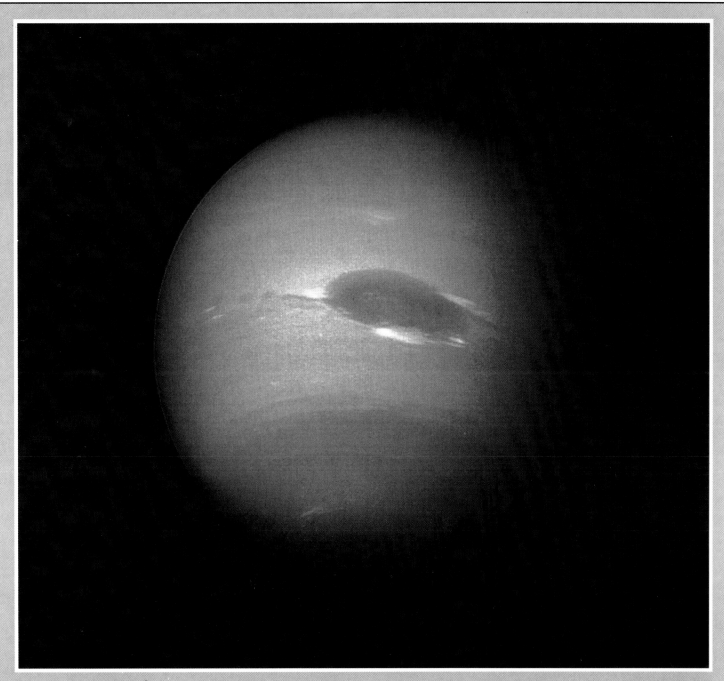

in Neptune's atmosphere is very much analogous to Jupiter's Great Red Spot, both in terms of location and its size relative to the planet. Like the Great Red Spot, the Great Dark Spot is an elliptical, stormlike feature that rotates in a counterclockwise direction and is probably located *above* the surrounding cloud tops. The Great Dark Spot is encircled by a constantly changing pattern of cirrus clouds and followed in its movement through Neptune's atmosphere by a string of much smaller elliptical storms. The cirrus clouds are composed of methane crystals and cling to the Great Dark Spot the way cirrus clouds cling to mountain tops on the Earth's islands. The Great Dark Spot completes a revolution of the planet in 18.3 hours, moving east to west.

To the south of the Great Dark Spot is another bright cloud which was observed by Voyager for the entire duration of the flyby and which was nicknamed 'Scooter' because it moves at a much faster relative speed than the Great Dark Spot, causing it to overtake the latter every five days. South of this feature is the Lesser Dark Spot with a permanent cloud bank situated over its center.

A New Generation of Interplanetary Spacecraft

During the late 1960s and the early 1970s, NASA conducted what amounted to a virtual interplanetary exploration launch blitz. A whole class of Mariner and Pioneer spacecraft were sent to every planet, except the outer three, and the Voyager project added two of those to the list of planets visited. Viking landed on the surface of Mars.

Budget cuts and delays held the whole notion of interplanetary exploration in abeyance throughout the 1980s. No interplanetary spacecraft would be launched until the eve of the 1990s. With the launch of the **Galileo** spacecraft on 18 October 1989, a new era in planetary exploration began.

Named for the Italian renaissance scientist who discovered Jupiter's major moons by using the first astronomical telescope, the Galileo mission to Jupiter was designed to study the planet's atmosphere, satellites and surrounding magnetosphere. This mission was the first to attempt direct measurements from an instrumented probe within Jupiter's atmosphere and the first to conduct *long-term* observations of the planet and its magnetosphere and satellites from orbit around Jupiter. It was to be the first orbiter and atmospheric probe for any of the outer planets.

The spacecraft passed Venus once and Earth twice using gravity assists from the planets to pick up enough speed to reach Jupiter. In December 1995, the Galileo atmospheric probe would conduct a brief, direct examination of Jupiter's atmosphere, while the larger part of the craft, the orbiter, would begin a 22-month, 10-orbit tour of major moons and the Jovian magnetosphere.

One year after launch, and still in an elliptical solar orbit, Galileo entered the asteroid belt, and two months later had its first asteroid encounter with the asteroid Gaspra. Although Galileo's rib-wrap (umbrella-type) antenna jammed partially when it was deployed, this did not effect the quality of the pictures that the spacecraft transmitted to Earth, only the speed of the transmissions. Gaspra is believed to be a fairly representative main-belt asteroid, about 10 miles across, and probably similar in composition to stony meteorites. Thirteen months later, the spacecraft completed its two-year elliptical orbit around the Sun and arrived back at Earth. The second flyby of Earth was designed to pump the orbit up to a larger ellipse, big enough to reach Jupiter.

Nine months after the final Earth flyby, Galileo had a second asteroid-observing opportunity. Ida is about 20 miles across. Like Gaspra, Ida is believed to represent the majority of main-belt asteroids in composition, though there are believed to be differences between the two.

Some two years after leaving Earth for the third time and five months before reaching Jupiter, Galileo's probe would separate from the orbiter, aimed precisely at its entry point in Jupiter's atmosphere. While the probe is still approaching Jupiter, the orbiter would have its first two satellite encounters. After passing within 20,000 miles of Europa, it would fly about 600 miles above Io's volcano-torn surface, 20 times closer than the closest flyby altitude of Voyager in 1979.

Opposite: **The burn of the IUS rocket carries the Galileo spacecraft away from the Space Shuttle.**

THIRD GENERATION NASA INTERPLANETARY EXPLORERS

- **Galileo** (1989): An orbital, atmosphere-probe-dropping mission to Jupiter and its major moons.
- **Mars Observer** (1992): The most sophisticated orbital spacecraft ever sent to another planet. The precursor to a manned landing on Mars.
- **Comet Rendezvous/Asteroid Flyby (CRAF)** (1995): A sophisticated six-year effort to study phenomena in the inner Solar System.
- **Cassini** (1996): An orbital, atmosphere-probe-dropping mission to Saturn and its moon Titan.

152 THE ILLUSTRATED GUIDE TO ASTRONOMY

The probe mission was designed with four phases: launch, cruise, coast and entry-descent. During launch and cruise, the probe would be carried by the orbiter and serviced by a common umbilical. The probe would be dormant during cruise, except for annual checkouts of spacecraft systems and instruments. During this period, the orbiter would provide the probe with electric power, commands, data transmission and some thermal control.

In the entry-descent phase, the probe would slam into Jupiter's atmosphere at 115,000 mph, fast enough to jet from Los Angeles to New York in 90 seconds. Deceleration to about Mach 1—the speed of sound—would take just a few minutes. At maximum deceleration as the craft slows from 115,000 mph to 100 mph, it would be hurtling against a force 350 times Earth's gravity. The incandescent shock wave ahead of the probe would be as bright as the Sun and reach searing temperatures of up to 28,000 degrees Fahrenheit. After the aerodynamic braking had slowed the probe, it would drop its heat shields and deploy its parachute. This would allow the probe to float down about 125 miles through the clouds, passing from a pressure of one-tenth that on Earth's surface to about 25 Earth atmospheres. About four minutes after the probe entry into Jupiter's atmosphere, the main parachute would open, and the probe would pass through the white cirrus clouds of ammonia crystals—the highest cloud deck. Beneath this ammonia layer probably lie the reddish-brown

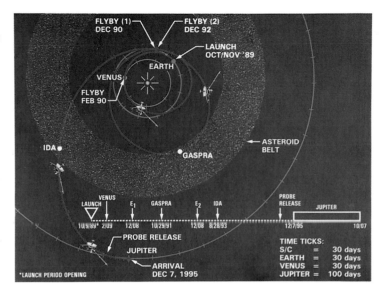

Clockwise from above: **A map of the trajectory of the Galileo mission; the Galileo probe; preliminary tests of the spacecraft in a solar thermal vacuum chamber; the STS launch; a diagram of the Galileo Orbiter.**

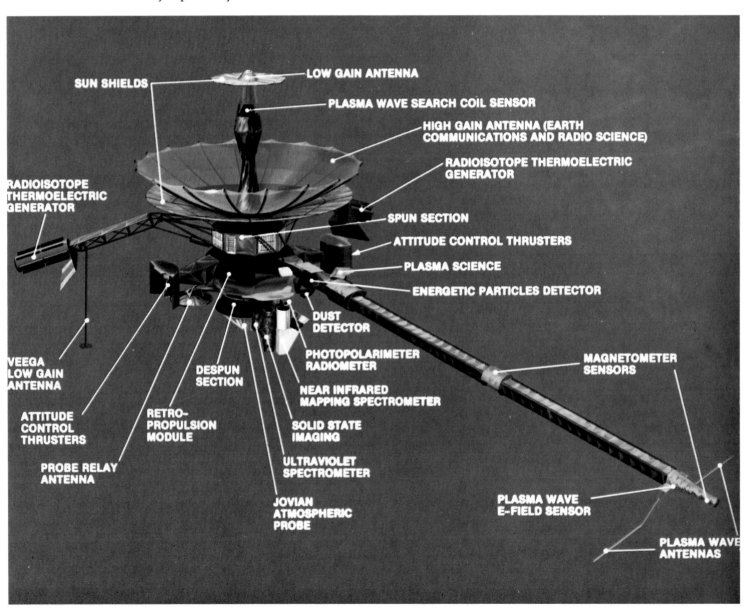

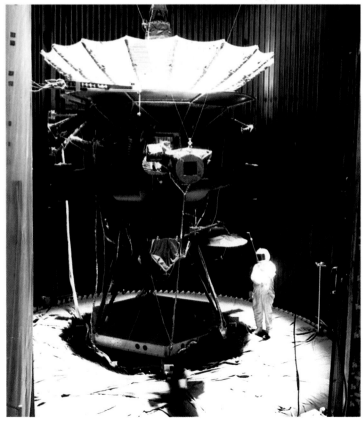

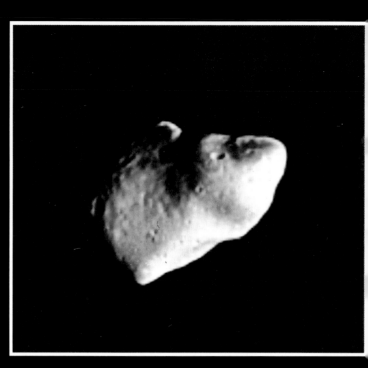

Above: Aboard the Space Shuttle *Atlantis* showing the Galileo spacecraft being deployed from the payload bay in 1989. *Below:* An artist's drawing of the Galileo Orbiter encountering Io.

Above: A Galileo photo of the asteroid Gaspra taken from a distance of 10,000 miles. *Right:* The Galileo probe enters the Jovian atmosphere in this artist's drawing.

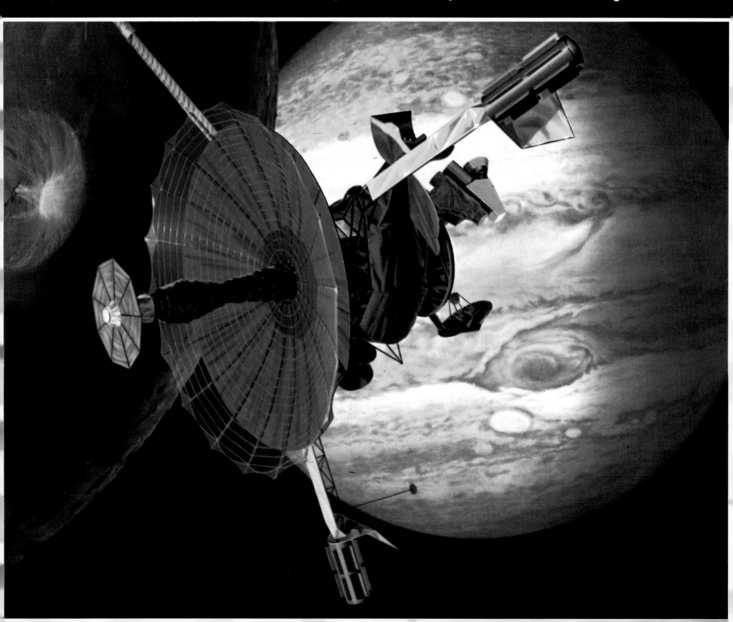

A NEW GENERATION OF INTERPLANETARY SPACECRAFT

clouds of ammonium hydrosulfides. Once past this layer, the probe would reach thick water clouds. Eventually, the probe would sink below these clouds, where rising pressure and temperature would destroy it. The probe's active life within Jupiter's atmosphere is predicted to be about 75 minutes.

NASA also continues exploration of Mars, started by the Mariner 4 spacecraft 28 years ago, with the **Mars Observer**, which was launched on 25 September 1992. 'Mars Observer will examine Mars much as Earth satellites now map our weather and resources,' said **Dr Wesley Huntress**, Director of NASA's Solar System Exploration Division in Washington, DC. 'It will give us a vast amount of geological and atmospheric information covering a full Martian year. At last we will know what Mars is actually like in all seasons, from the ground up, pole to pole.'

'Mars Observer will tell us far more about Mars than we've learned from all previous missions to date,' said **David Evans**, Project Manager at NASA's Jet Propulsion Laboratory. 'We want to put together a global portrait of Mars as it exists today and, with that information, we can begin to understand the history of Mars. By studying the evolution of Mars, as well as Venus, we hope to develop a better understanding as to what is now happening to planet Earth. As we look even further into the future, this survey will be used to guide future expeditions to Mars. The first humans to set foot on the planet will certainly use Mars Observer maps and rely on its geologic and climatic data.'

On 19 August 1993, Mars Observer arrived in the vicinity of Mars, and as it approaches the planet, onboard rocket engines would slow its speed and allow the gravity of Mars to capture it in orbit around the planet. Mars Observer would first enter a highly elliptical orbit. Then, over a period of four months, onboard rocket thrusters would gradually move the spacecraft into a nearly circular orbit inclined 93 degrees to the planet's equator, 200 miles above the Martian surface. In this orbit, the spacecraft flew near the Martian poles.

Mars Observer was designed to provide scientists with an orbital platform from which the entire Martian surface and atmosphere could be examined and mapped during the mapping cycle starting on 13 January 1994. The measurements would be collected daily from the low-altitude polar orbit, over the course of one complete Martian year—the equivalent of 687 Earth days.

Left: The Cassini spacecraft arrives at Titan, Saturn's largest moon, in this artist's drawing. *Above:* The Mars Observer spacecraft scans the surface of the red planet from its orbit in this artist's drawing. *Right:* In this artist's drawing, the Comet Rendezvous Asteroid Flyby (CRAF) ejects a penetrator toward the nucleus of a comet. It is the first mission of the Mariner Mark II series of spacecraft.

The New Orbiting Observatories

A NASA astrophysics spacecraft, the **Cosmic Background Explorer (COBE)**, was launched on 18 November 1989. In 1990, it undertook a survey of the entire sky in infrared and microwave radiation, making unprecedented measurements of background radiation. In April 1990, COBE sent back to Earth a clear infrared view of the center of our own galaxy, which is usually obscured from view in visible light by interstellar dust.

Preliminary results from COBE were in accord with the predictions of the Big Bang theory, a theory which traces the origin of the universe to a primordial explosion some 15 billion years ago. The universe today shows that sometime after the Big Bang, additional release(s) of energy must have occurred. COBE's data severely limited the magnitude and character of such a release. COBE data indicated a smooth, uniform Big Bang. However, small deviations from a blackbody spectrum—the characteristic signature of radiation from an opaque object of uniform temperature—revealed energetic processes in the early universe.

In October 1990, COBE took its first image of the entire Milky Way Galaxy in a wavelength that revealed the dust from which planets and galaxies were formed. The 1990 image, taken in far infrared wavelength by the **Diffuse Infrared Background Experiment**, showed radiation from cold interstellar dust. In April 1991, COBE scientists released an image take by the same instrument in the near infrared wavelength, which revealed millions of stars in this galaxy. A comparison of these images showed the difference in the spatial distribution of the galaxy's stellar and interstellar components.

In January 1991, for the first time, astronomers mapped the distribution of nitrogen throughout our galaxy. The new observations were taken by the **Far Infrared Absolute Spectrophotometer** on the COBE. This all-sky survey, along with additional maps of carbon and dust, provided quantitative information that may one day enable scientists to better understand the heating and cooling processes that take place throughout the Milky Way. These accomplishments were reported at a meeting of the American Astronomical Society in Philadelphia by members of the COBE science team.

The scientists presented images that showed the locations in the galaxy of ionized nitrogen. The nitrogen map of the Milky Way is at a wavelength of 205 micrometers, and theirs was the first detection of this important spectral line. Their carbon map was produced at 158 micrometers and the dust map at 205 micrometer wavelengths. Five months of data were used to produce the maps.

'Before COBE, it was not possible to map the whole galaxy in this way, although these atomic emissions are the dominant way in which interstellar gas cools,' said COBE scientist **Dr John C Mather**, who added that COBE's unique capabilities permit such an all-sky measurement unencumbered by atmospheric and instrument emission.

This data showed that carbon and nitrogen atoms—some of the key building blocks of life—were extremely widespread in the thin gas that fills the space between the stars. These

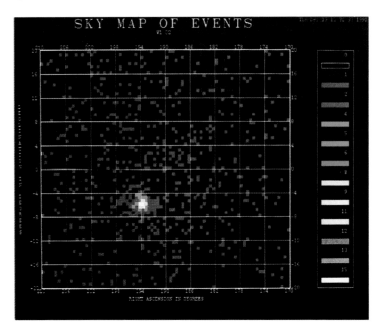

Below: **A digitized, false-color gamma-ray sky map made by the Energetic Gamma Ray Experiment Telescope (EGRET) aboard the Compton Observatory *(opposite)*. The bright spot is quasar 3C 279.**

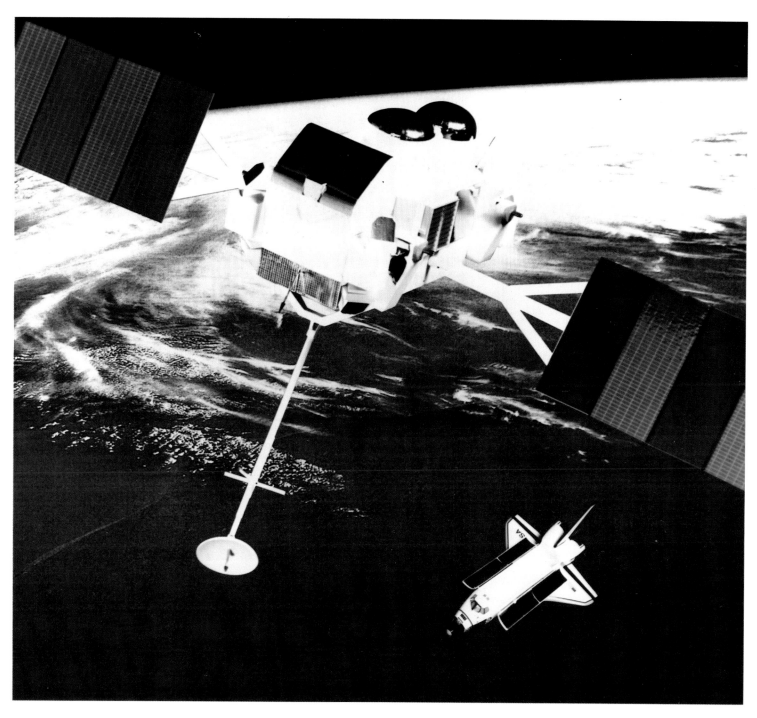

atoms are created inside stars by nuclear reactions and then released back into space by stellar winds or explosions at the end of their lives. The data also confirmed theories that the mixture of gas and dust in our galaxy is heated by starlight striking dust grains and cooled by the carbon and nitrogen emissions. The greatest concentrations of the atoms and dust were found in the plane of the galaxy.

NASA's **Gamma Ray Observatory (GRO)** was deployed from the Space Shuttle Orbiter *Atlantis* on 7 April 1991, and maneuvered to point at its first scientific target of opportunity on 7 June 1991 at 7:45 pm EDT. On 23 September 1991, NASA announced that the GRO would be renamed in honor of Nobel Prize-winning American physicist **Dr Arthur Holly Compton (1892-1962)**. The new official name given to the 17-ton orbiting spacecraft was the Arthur Holly Compton Gamma Ray Observatory, or Compton Observatory. Compton is known for his ground-breaking series of experiments on the interaction of high-energy radiation and matter that demonstrated the wave/particle duality of nature. His findings played a key role in the development of modern physics. In the late 1930s, Compton conducted comprehensive studies of cosmic rays. Interactions of cosmic rays with interstellar gas are an important source of gamma rays that the Compton Observatory is studying. Compton's work provided significant clues to our present understanding of many of the basic physical processes that create gamma radiation.

The 17-ton GRO was repositioned by controllers at the Goddard Space Flight Center at Greenbelt, Maryland, to gather data from two X-class solar flares, the largest and most powerful type of solar flare. Solar flares are temporary outbursts of intense solar radiation that have been observed blasting hot loops of gas more than 430,000 miles into space. These high energy outbursts have been known to disrupt the Earth's magnetic field and cause interference with communications equipment and electrical power distribution systems.

While much is known about the composition and magnitude of solar flares, surprisingly little is known about the thermonuclear processes of these dynamic solar phenomena. The decision to observe the sun was made by project scientists and engineers around noon on 7 June 1991. The flight opera-

Left: This red light photograph of the Spiral Galaxy M81 in the constellation Ursa Major was made during the ASTRO-1 mission of the Space Shuttle *Columbia* by a 36-inch telescope at Kitt Peak National Observatory. The galaxy is about 12 million light years from Earth. *Left below:* The same galaxy as photographed by the Ultraviolet Imaging Telescope (UIT), a 15-inch telescope built at Goddard and one of four telescopes that flew on the ASTRO-1 mission 2-11 December 1990. The bright spots in the curved arms of the galaxy are concentrations of very young, hot stars. *Right:* Omega Centauri within our own Milky Way galaxy was photographed for the first time in ultraviolet light by the UIT on the ASTRO-1 mission. Analyses of UIT star cluster photos are expected to add to astronomers' understanding of stellar evolution.

tions team was able to complete the maneuver in nine hours. The normal time for this type of maneuver is usually about 36 hours. The fast action of the team gained 23 additional hours of observing time, allowing GRO to capture data on a solar flare the following day which otherwise would have been impossible.

In July 1991, the Goddard Space Flight Center's **Energetic Gamma Ray Experiment Telescope (EGRET)**, one of four instruments aboard the Compton GRO, detected 'the most distant, and by far the most luminous, gamma ray source ever seen,' according to EGRET principal investigator **Dr Carl Fichtel**. The EGRET team, led by Fichtel, reported in a telegram to the International Astronomical Union at Cambridge, Massachusetts, that a source of intense localized gamma radiation had been detected between 15 and 28 June. The source of the radiation was identified with the variable Quasar 3C279, located in the constellation Virgo, approximately seven billion light years from Earth.

'Quasar 3C279 is a variable quasar, meaning that its intensity changes over time,' Fichtel explained. At the intensity the scientists measured, this source should have been visible to two previous gamma ray missions—NASA's Small Astronomy Satellite and the European Celestial Observation Satellite. In their telegram, the EGRET team stated that 'neither reported a detection during 1972 to 1973 or between 1975 and 1982, respectively.'

In January 1992, the Compton GRO found three new gamma ray quasars, detected more than 200 cosmic gamma ray bursts and captured the best ever observation of the glow of gamma radiation from the disk of the Milky Way Galaxy. Dr Fichtel stated that the EGRET appeared to have detected 'still more distant and very luminous gamma ray sources, even more distant than the massive quasar 3C279.'

The EGRET team reported three sources of intense, localized gamma radiation, quasars Q0208-512, 4C38.41 and PKS0528+134, detected between 16 May and 18 September 1991, located in the constellations of Eridanus, Hercules and near the Crab Nebula, approximately 10 to 20 billion light years from Earth. 'The sources are emitting an extraordinary flux of gamma rays, each gamma ray photon with an energy greater than 100 million electron volts. In contrast, a visible light photon has an energy of only a few electron volts, and an X-ray photon has an energy of 1000 electron volts' Fichtel said, 'The luminosity or total energy emitted by these sources is approximately 10 to 100 million times the total gamma ray emission of the Milky Way Galaxy.'

During the first part of April 1992, the Compton GRO made the first detection of high-energy gamma rays from a class of active galaxies similar to quasars. The observations, made by the EGRET, suggest that high-energy gamma radiation provided a substantial contribution to the objects' overall luminosity. These active galaxies are called **BL Lacertae** objects, a type of 'quasar-like' object that emits vast but varying amounts of energy. The candidate objects were in the constellations Ursa Major, Pictor and Camlopardalis. They were designated as MK 421, 0537-441 and 0716+714, respectively.

These new gamma ray results supported the hypothesis that objects, like quasars, may be powered by supermassive black holes. The detection of these high-energy gamma rays also provided another piece of evidence suggesting their similarity to quasars and added important new insight into understanding the nature of BL Lacertae objects. The EGRET science team previously reported detection of high energy gamma rays from six quasars.

On 23 April 1992, scientists announced at the American Physical Society's meeting, held in Washington, DC, that they had detected the long-sought variations within the glow from the Big Bang, using COBE. This detection was a major milestone in a 25-year search, and their findings supported theories explaining how the initial expansion happened.

These variations showed up as temperature fluctuations in the sky, revealed by statistical analysis of maps made by the **Differential Microwave Radiometers (DMR)** on the COBE satellite. The fluctuations were extremely faint, only about 30 millionths of a degree warmer or cooler than the rest of the sky, which is itself very cold—only 2.73 degrees above absolute zero.

The Big Bang theory was initially suggested because it explained why distant galaxies are receding from Earth at enormous speeds, as though all galaxies started moving away from the same location a long time ago. The theory also predicts the existence of cosmic background radiation—the glow left over from the explosion itself. The Big Bang theory received its strongest confirmation when this radiation was discovered in 1974 by **Arno Penzias** and **Robert Wilson**, who later won the Nobel Prize for this discovery.

Although the Big Bang theory is widely accepted, there have been several unresolved mysteries. How could all of the matter and energy in the universe become so evenly mixed in the instant following the Big Bang? How could this evenly distributed matter then break up spontaneously into objects of all sizes, such as galaxies and clusters of galaxies? The temperature variations seen by COBE help to resolve these mysteries.

'The COBE receivers mapped the sky as it would appear if our eyes could see microwaves at the wavelengths 3.3, 5.7 and 9.6 mm, which is about 10,000 times longer than the wavelength of ordinary light,' explained **Dr George Smoot** at the

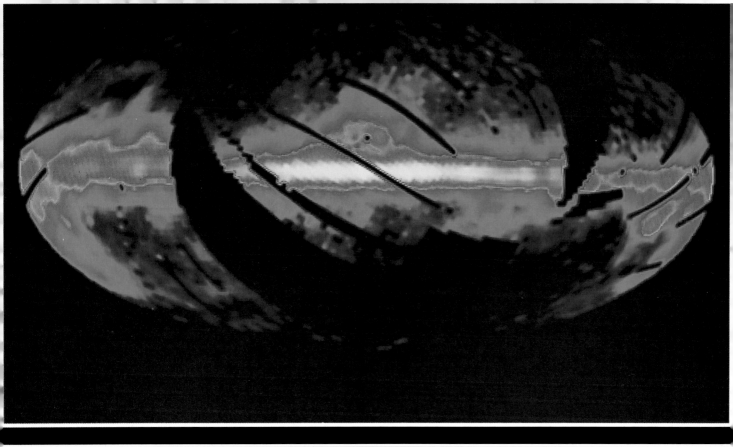

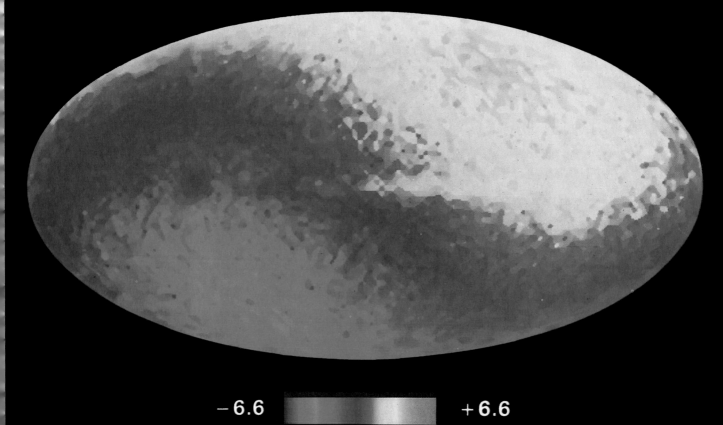

COBE Differential Microwave Radiometers
FULL SKY MICROWAVE MAP
53 GHz   5.7 mm

−6.6   +6.6
mK

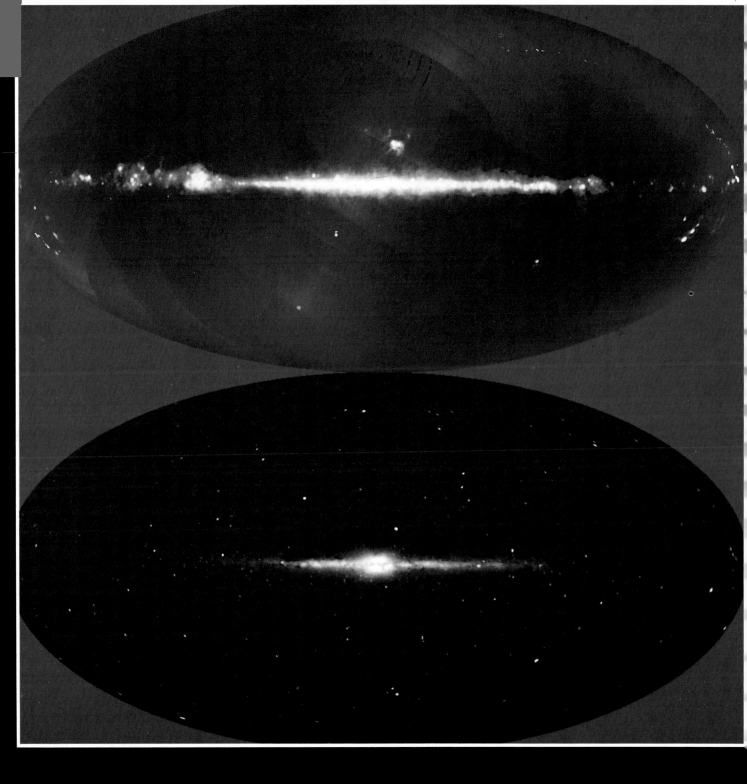

Left above: This all-sky map constructed from five months of data obtained by the Far Infrared Explorer aboard NASA's Cosmic Background Explorer (COBE) shows the emission from dust grains in the space between the stars at wavelengths near 205 micrometers. The COBE offers an important key to understanding heating and cooling of the interstellar material in our galaxy. The yellow areas are the most intense. The plane of the Milky Way galaxy is horizontal across the middle with the galactic center at the center. *Left:* This is a representation of a full-sky map constructed from six months of data obtained by the Differential Microwave Radiometers (DMR) instrument aboard the COBE. The smooth variation between the hot areas (pink) and cold areas (light blue) on opposite sides of the sky is a very faint signal. *Above:* These images of the infrared sky were obtained by the Diffuse Infrared Background Experiment (DIRBE) aboard the COBE. Shown are both the thin disk and central bulge populations of stars in our spiral galaxy. The image is redder in directions where there is more dust between the stars absorbing starlight from distant stars. The data is studied for evidence of a faint uniform infrared background of residual radiation from the first stars and galaxies formed following the Big Bang.

University of California, Berkeley, the leader of the team that made this discovery. 'Most of the energy received from the sky at these wavelengths is from the cosmic background radiation of the Big Bang, but it is extremely faint by human standards. Hundreds of millions of measurements were made by the DMR over the course of a year, and then combined to make pictures of the sky. Making sure all the measurements were combined correctly required exquisitely careful computer analysis,' Smoot explained.

Another COBE scientist, **Dr Charles Bennett** of the Goddard Space Flight Center, explained that a major challenge for the team has been to distinguish the Big Bang signals from those coming from our own Milky Way Galaxy. He pointed out that 'the Milky Way emits microwaves that appear mostly concentrated in a narrow zone around the sky. We compared the signals at different positions and at different wavelengths to separate the radiation of the Big Bang from that of the Milky Way Galaxy.'

In May 1992, the Compton GRO solved a 20-year-old mystery about the nature of Geminga, one of the brightest emitters of high-energy gamma rays in the sky. Scientists were unclear about the source of Geminga's power and why it shone brightly in gamma rays. Using data from two different spacecraft, scientists now know that the power plant in Geminga is a rotating, 300,000-year-old neutron star. **Dr Jules Halpern** of Columbia University and **Dr Stephen Holt** of NASA's Goddard Space Flight Center reported in *Nature* that they had observed X-ray pulsations from Geminga using data from the **German/American Roentgen Satellite (ROSAT)**. These observations firmly establish Geminga as a close cousin of the Crab and Vela nebulae, which also have pulsating neutron stars at their cores. The rotating neutron star produces focused beams of radiation, much like a periodic flashing or pulsating lighthouse beacon.

In a companion paper, an investigative team, led by Goddard's **Dr David Bertsch**, confirmed the pulsations using gamma ray data from EGRET on the Compton GRO, and further estimated the age of Geminga as 300,000 years. 'With this discovery, we consider the mystery of Geminga largely solved,' said Halpern. Geminga was discovered in 1972 in the first high-energy gamma ray survey of the sky, which was conducted with NASA's **Small Astronomical Satellite 2**. Located only a few degrees from the Crab Nebula, it is one of the brightest X-ray and gamma ray sources in the sky.

Geminga is even brighter than the Crab in gamma rays. Because it has no obvious optical or X-ray counterpart like the Crab Nebula, for example, it was given the name Geminga, which means 'it isn't there' in the Milanese Italian dialect. With virtually all of its power emitted in gamma rays, its nature had been a true mystery for a long time. In late 1980s, a

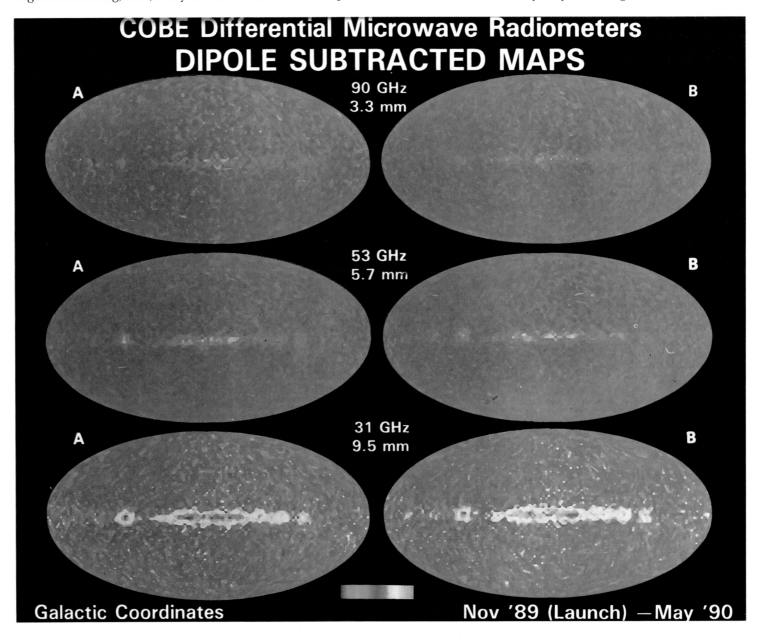

very weak X-ray source was suggested as a potential identification for Geminga on the basis of approximate positional coincidence. Using the ROSAT observatory, Halpern and Holt discovered that the X-ray intensity of this source was modulated with a period of 237 milliseconds, more than seven times longer than the period of the pulsar in the Crab Nebula and three times longer than the pulsar in the Vela Nebula.

The pulsations were the key to understanding the nature of Geminga. The Crab and Vela nebulae are supernova remnants powered by converting the kinetic energy of pulsars at their centers into radiation that generally includes the whole electromagnetic spectrum. Why Geminga is powerful only in gamma rays has not been determined, but the discovery of the new pulsar firmly established Geminga as a close cousin of the Crab and Vela nebulae.

Bertsch and the EGRET team confirmed that the gamma ray emission from the source modulated with the same period. Their observations also allowed the scientists to measure the rate at which the neutron star was slowing, providing an estimate of the age of the supernova that created the neutron star. This technique determined the age of Crab and Vela to be 1000 and 10,000 years, respectively. This discovery not only explained the nature of Geminga, but suggested that many of the remaining unidentified gamma ray sources in the Milky Way Galaxy also may be neutron stars. Although nearly all pulsars are discovered because of their strong radio signals, Geminga was apparently silent in the radio band. It is possible that Geminga is not unique in this regard.

The **Extreme Ultraviolet Explorer Spacecraft (EUVE)** was launched into Earth orbit on 7 June 1992 to search the spectrum between visible light and X-rays. By October, a powerful, exotic object two billion light years beyond the Milky Way Galaxy was observed by astronomers using a new NASA spacecraft designed to detect radiation in the little-explored, extreme ultraviolet portion of the electromagnetic spectrum.

'Twenty years ago, no one would have believed you could see out of the Solar System at EUV wavelengths. But now—for the first time—we actually have obtained an EUV spectrum for an object beyond our galaxy,' said **Dr Ed Weiler**, Chief of NASA's Ultraviolet and Visible Astrophysics Branch.

Observation of the EUV spectrum, both inside and out of the Milky Way Galaxy, is often blocked by gas and dust in interstellar space. However, the distribution of the gas and dust is uneven, which allows the EUVE telescopes to see distant sources of radiation.

One of the new EUV sources detected by the EUVE early in the mission was the corona of a star much like the Sun, located about 16 light years away from Earth. A white dwarf companion seven arc minutes away also appears in the EUV image. On 8-9 July 1992, an outburst was observed from a cataclysmic variable, RE 1938-461, a closely orbiting pair of stars in which gravitational forces pull matter from the outermost layers of one star onto the surface of the other, a white dwarf companion. The hot, compressed stellar material generates an explosive burst of EUV radiation as it falls into the deep gravitational field of the white dwarf.

Other explosive events are flares on stars. These are unpredictable, giant versions of eruptions known to occur on a much smaller scale on the Sun. The EUVE caught two such events on the red dwarf stars AT Mic and AU Mic. It was announced on 7 January 1993 that the Big Bang theory passed its toughest test yet with the results reported from COBE. Precise measurements made by COBE's **Far Infrared Abso-**

Opposite: **The six all-sky maps were constructed from data from the DMR aboard the COBE. The dipole anistropy, that is the motion of the Solar System relative to the distant matter, has been subtracted. The Milky Way galaxy can be seen in all the maps. Microwave emissions are strongest at 31.5 GHz and weaker, but still present, at 53 and 90 GHz.** *Above:* **This image taken by NASA's Extreme Ultraviolet Explorer (EUVE) shows two sources of extreme ultraviolet radiation, HR 6094** *(top)* **and WD1620-391** *(bottom),* **a binary star system in the constellation Scorpio located 48 light years away from Earth.**

lute Spectrophotometer (FIRAS) of the afterglow from the Big Bang—the primeval explosion that began the universe approximately 15 billion years ago—show that 99.97 percent of the early radiant energy of the universe was released within the first year after the Big Bang itself. 'Radiant energy' is energy emitted in any form of light, from X-rays and gamma rays to visible and infrared light or even radio waves. COBE's FIRAS was designed to received the microwave and infrared energy from the Big Bang.

'The Big Bang theory comes out a winner,' said COBE project scientist and FIRAS principal investigator **Dr John C Mather** of NASA's Goddard Space Flight Center. 'This is the ultimate in tracing one's cosmic roots.'

All theories that attempt to explain the origin of large-scale structures seen in the universe today now must conform to the constraints imposed by these measurements. This includes theories that postulate large amounts of energy released by such things as black holes, exploding supermassive stars or the decay of unstable elementary particles. In other words, there were not a lot of 'little bangs,' as suggested by some theories.

Yesterday and Tomorrow

Placing a telescope in space has been a dream of astronomers for decades, dating back to before space flight was a fact. There, far above the distorting effects of the Earth's atmosphere, astronomers would have an unimpaired view of the entire universe. NASA sought to realize this dream through the **Edwin P Hubble Space Telescope (HST)**, a national orbiting observatory which was placed into orbit 320 nautical miles above the Earth on 24 April 1990.

The Hubble Space Telescope is a 94.5-inch Ritchey-Chretien telescope, named in honor of American astronomer **Edwin P Hubble**, who made vital contributions to the understanding of galaxies and the universe through his work earlier this century. Despite spherical abberations in the telescope's main mirror, the telescope's high vantage point allows it to see farther and with greater clarity than any astronomical instrument ever built. Besides 'seeing' better it will detect more portions of the electromagnetic spectrum, such as ultraviolet light which is absorbed by the atmosphere before reaching the ground.

The heart of the Space Telescope is the **Optical Telescope Assembly (OTA)**. The major segments of the OTA are the 94.5-inch primary mirror, a secondary mirror of 12 inches, and the OTA's support structure. Light entering the Space Telescope is reflected off the primary mirror to the secondary mirror, 16 feet away. The secondary mirror sends the light through a hole in the center of the large mirror, back to the scientific instruments. Providing all the essential systems to keep the Hubble Telescope operating in the hostile environment of space is the function of the Support Systems Module. The SSM also directs communications, commands, power, and fine pointing control for the telescope. Collectively, the SSM consists of the light shield on the front end of the telescope; the equipment section, with the main spacecraft electronics equipment; and the aft shroud, which contains the scientific instruments.

The **Fine Guidance Sensors (FGS)** feed roll, pitch and yaw information to the telescope's attitude control system. The pointing capability of the Hubble Telescope provided by the FGS is so precise it is often called a sixth scientific instrument. To point the telescope, the FGS must identify the position of specified stars. This pointing data can be used to calibrate space-distance relationships throughout the universe.

The Hubble Space Telescope carries five scientific instruments which are replaceable and serviceable on orbit. Four of the instruments, about the size of telephone booths, are located in the aft shroud, behind the primary mirror. They receive light directly from the secondary mirror. The fifth instrument, the **Wide Field/Planetary Camera (WFPC)**, is located on the circumference of the telescope and uses a pick-off mirror system.

The **Faint Object Camera (FOC)** does exactly what its name implies, observes faint objects. It does this by taking very low light levels and electronically intensifying the images. Objects as faint as 28th or 29th magnitude (the higher the magnitude, the fainter the object) should be observed by the FOC. By comparison, Earth-based telescopes can see to about 24th magnitude. Likely targets for the instrument are the search for extra-solar planets, variable brightness stars, and in its spectrographic mode, the center of galaxies suspected of concealing black holes.

The FOC measures the chemical composition of very faint objects. Visible light contains information used to determine the chemical elements which make up the light source. Special gratings and filters allow the FOC to make spectral expo-

Below: **The Hubble Space Telescope.** *Opposite:* **An artist's drawing of a service shuttle approaching the HST.**

sures which not only reveal information about the make-up of a light source but also about its temperature, motion and physical characteristics. This instrument will study the spectra of objects in the ultraviolet and visible wavelengths. Particular targets of interest are quasars, comets and galaxies.

The Office of Space Science and Application at NASA Headquarters is responsible for overall program management, financial and scheduling provisions, and the science policy development and direction. Marshall Space Flight Center in Huntsville, Alabama is responsible for the development and operation of the Space Telescope system as the 'lead' NASA Center. In Greenbelt, Maryland, the Goddard Space Flight Center is managing the scientific instruments, mission operations, and data management. It is also charged with monitoring the Space Telescope Science Institute.

The Space Telescope Science Institute is operated by AURA, the Association of Universities for Research in Astronomy. The Institute is located on the Homewood Campus of Johns Hopkins University in Baltimore, Maryland. It is the job of the Institute to determine the observational program of the Space Telescope while in orbit, insuring that the observatory will be used to its maximum advantage.

The Hubble Space Telescope epitomized 1990's highs and lows before finishing the year on an optimistic note. After years of anticipation, the HST was launched aboard the Space Shuttle Orbiter *Discovery* in April 1990. By orbiting above the Earth's atmosphere, which distorts astronomers' observations, Hubble was supposed to be able to see more clearly than any other telescope ever had, gathering data on the origins of the universe. Its very first optical-engineering test returned a valuable science observation, resolving the star cluster 30 Doradus three to four times better than the best ground-based observation.

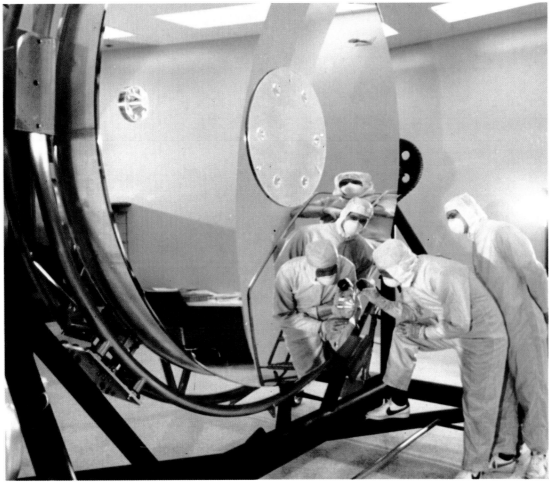

Left above: **Construction of the HST in Sunnyvale, California.** *Left:* **The infamous HST mirror.** *Right:* **A diagram of the Hubble Space Telescope.**

The elation over Hubble was deflated by the discovery in June 1990 that the telescope's optical system was affected by a spherical aberration, a misshaping of the primary mirror that prevented the telescope from focusing light to a single, precise point. As a result, the telescope could not see very faint objects or distinguish faint objects in a crowded field.

Dr Lennard A Fisk, the associate administrator for Space Science and Applications, appointed a review board, chaired by **Dr Lew Allen**, the former director of the Jet Propulsion Laboratory, to investigate how the aberration had occurred and why it had not been detected before launch. After five months, the board concluded that a key device used to test the shape of the primary mirror had been assembled incorrectly. As a result, the mirror was improperly ground by two to three microns (thousandths of a millimeter) at its edge.

The board's report faulted the contractor, Perkin-Elmer (now Hughes Danbury Optical Systems) and NASA for quality control practices that had allowed the error to remain uncorrected. The board also concluded that mechanisms were in place to prevent similar problems from affecting future systems, such as the Advanced X-Ray Astrophysics Facility.

Spherical aberration notwithstanding, the HST proved itself an exemplary observatory, still capable of seeing objects in visible light much more clearly than ground-based telescopes and making extraordinary observations in the ultraviolet wavelengths, which are blocked from the surface by the Earth's atmosphere. The WFPC observed a jet of material streaming away from the Orion Nebula with unprecedented clarity, offering insights into this region of young stars. The Faint Object Camera, built by the European Space Agency, returned the clearest image yet of Pluto, and its moon, Charon. Most dramatically, the WFPC took several hundred pictures as the white spots on Saturn grew into an immense storm that spread around the planet's equator.

The discovery of the spherical aberration gave new emphasis to a previously planned HST servicing mission when astronauts will replace the WFPC with its backup unit, WFPC-2, which will be modified to compensate for the spherical aberration. Once this is accomplished, NASA astronomers are confident HST will achieve most of its major scientific objectives.

On 17 August 1990, despite its infirmity, HST provided a new, remarkably detailed view of the core of a galaxy, which is

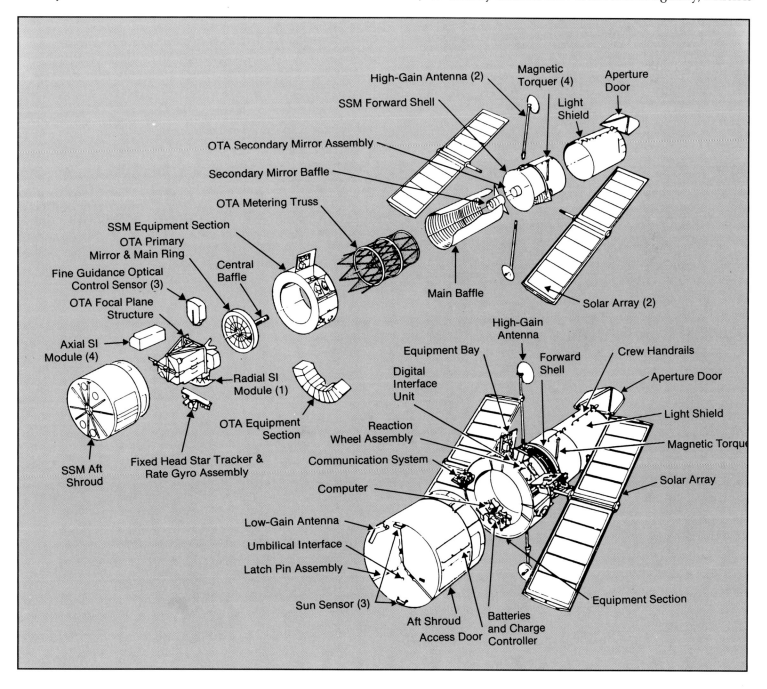

Left: The launch of the five-day Space Shuttle *Discovery* STS-31 mission. *Above:* An artist's conception of the deployment of the Hubble Space Telescope. *Below:* The HST is released from its moorings in the cargo bay by the handlike end effector of the remote manipulating system. *Right:* The HST is free of *Discovery* and operating under its own power.

These pages: **These sharp images of Saturn and its ring system *(above),* Mars *(right above)* and Jupiter *(right below)* taken by HST's WFPC, surpass any taken by ground-based telescopes.**

40 million lightyears away, more than half way to the great Virgo cluster of galaxies. The image revealed that stars were much more tightly concentrated at the center of the galaxy than was previously expected. Since the galaxy, catalogued as NGC 7457, was assumed to be a 'typical' galaxy, the preliminary findings suggested that the nuclei of normal galaxies may be more densely packed with stars than previously thought.

'The images of NGC 7457 showed emphatically that research on nuclei of galaxies can still be done,' said **Tod Lauer** of the Wide Field and Planetary Camera imaging team. 'We've never been able to study any galaxy outside of our local group, our 'neighborhood' of about two dozen galaxies, at this resolution before.' The centers of galaxies are extremely interesting to astronomers because the centers are at the heart of the violent processes that give rise to cosmic jets, quasars, Seyfert galaxies and other mysterious energetic behavior.

NGC 7457 was a quiescent galaxy picked for its 'normality' as an early target for assessing the science performance of HST. The resulting images showed, to the surprise of astronomers, that an exceptionally bright and compact core was embedded in the diffuse background of the rest of the galaxy. Based on this image, the stars in the nucleus of NGC 7457 were crowded together at least 30,000 times more densely than those stars seen within our own galactic neighborhood. This extraordinarily high stellar density exceeded earlier estimates from ground-based observation of NGC 7457 by a factor of 400.

It is far from clear whether or not a massive black hole is at the center of NGC 7457, since the images alone did not provide the answer. However, this data suggested that NGC 7457 was an excellent place to use the HST's spectrographs to measure how much mass was concentrated at the center of the galaxy.

Observations made with the European Space Agency's Faint Object Camera on 23-24 August 1990 provided, with unprecedented sharpness and clarity down to .1 arc second, an intriguing view of the supernova and its surrounding shell of stellar material. The image, taken in visible light, revealed the details of the circumstellar shell, whose characteristics previously had been suggested by ground-based observations and data from the International Ultraviolet Explorer satellite.

The visible light image clearly showed an elliptical, luminescent ring of gas about 1.3 light years across, surrounding the still-glowing center of the 1987 explosion. The ring was a relic of the hydrogen-rich stellar envelope that was ejected in the form of a gentle 'stellar wind' by the progenitor, which was a red supergiant star that existed an estimated 10,000 years before the explosion took place. On 16 January 1991 astronomers working with HST's Goddard High Resolution Spectrograph reported on what they called the best spectrograms ever obtained of Melnick 42, a very massive star in a galaxy 170,000 light years from Earth. The report was presented to the meeting of the American Astronomical Society in Philadelphia by a team led by **Dr Sally Heap** of NASA's Goddard Space Flight Center. Dr Heap said that preliminary analysis of the spectrograms showed that Melnick 42 was between 80 and 100 times more massive than the Sun, making it one of the most massive known stars. Further, the analysis revealed that Melnick 42 was shedding its hot gases at a furious rate in a so-called 'stellar wind' that strips the star of an amount of gas equal in mass to the Sun every 100,000 years.

Dr Heap explained that Melnick 42 is a hot, young supergiant star in the Large Magellanic Cloud (LMC), a galaxy neighboring the Milky Way. The star may be only 2 million years old, compared with the 4.6-billion-year age of the Earth. Melnick 42 has a surface temperature of about 86,000 degrees Fahrenheit, or eight times hotter than the Sun. According to present theory, Melnick 42 will explode as a supernova within the next few million years, while the Sun will continue to shine for several billion years. Melnick 42 is more than a million times brighter than the Sun.

Astronomers reported on 13 January 1992 that recent ultraviolet observations with the HST suggested that what were thought to be randomly distributed, nearly primordial clouds of hydrogen, may actually be associated with galaxies or clusters of galaxies. 'This is a revolutionary finding, if supported by future observations,' said **Dr John Bahcall** of the Institute for Advanced Study in Princeton, New Jersey. 'We would have never thought of looking for this explanation if it hadn't kicked us in the face.'

Using a unique capability of the HST, astronomers also announced that they had detected the rare element boron in an ancient star. This element may be 'fossil' evidence of energetic events which accompanied the birth of the Milky Way Galaxy. An alternative possibility is that this rare element may be even older, dating from the birth of our universe. If so, then the HST findings could force some modification in theories of the Big Bang itself.

The light from boron only appears in the ultraviolet part of the spectrum, and so does not penetrate Earth's atmosphere. 'That's why no one was able to make this discovery before,' said **Dr Douglas Duncan** of the Space Telescope Science Institute in Baltimore. 'Having a powerful telescope high above Earth's absorbing atmosphere has given scientists a new window on the universe. This was always considered to be one of the most important reasons for building the Space Telescope.'

Meanwhile, another team of scientists using HST made the most precise measurement to date of the percent of heavy hydrogen in space, which better determines the physical conditions present in the theorized Big Bang at the origin of the universe. The team's findings were presented at a press conference at the American Astronomical Society's semi-annual meeting in Atlanta. The group was headed by **Jeffrey Linsky**, a National Institute of Standards and Technology (NIST) astronomer and a fellow of the Joint Institute for Laboratory Astrophysics (JILA), a collaboration of NIST and the University of Colorado at Boulder.

Euterium, also called heavy hydrogen, differs from ordinary hydrogen by having one neutron in addition to one proton in its nucleus. A measurement of the ratio of deuterium to ordinary hydrogen provides a critical test of conditions in the universe at the time of the Big Bang because it is believed that essentially all of the deuterium now present was created at that time.

Linksy and his collaborators used the HST's Goddard High Resolution Spectrograph to observe the nearby star Capella, the brightest star in the constellation Auriga and the universe's sixth brightest star. With the highest spectral resolution possible on the HST, the scientists observed the far ultra-

violet light of the star where the strongest absorption by normal hydrogen and deuterium occurs. Their careful analysis indicated that the ratio of deuterium to hydrogen in space is 15 parts per million, with an uncertainty of less than 10 percent. If further research finds no evidence for large amounts of 'missing matter,' then their deuterium measurements strengthen the theory that the universe will expand forever. If true, the universe had a brilliant beginning but will have no end.

The Hubble Space Telescope has also provided intriguing new clues to cataclysmic events in the history of the peculiar galaxy NGC 1275, located approximately 200 million light years from Earth. Astronomers have discovered about 50 bright objects at the center of the galaxy which appear to be young, massive, globular star clusters. This discovery was surprising because most globular clusters are among the oldest objects in the universe. In fact, these clusters are used as a bench mark for estimating the age of the universe. 'Such objects have never before been seen,' said Dr Jon Holtzman of Lowell Observatory at Flagstaff, Arizona, who led the observing team that made the finding.

Globular clusters are dense, spherical collections of stars, containing 100,000 to 10 million stars packed in a region only 100 light years in diameter. More than 100 globular clusters orbit the Milky Way in a diffuse swarm. The brightest of these appear as 'fuzzy' stars to the naked eye.

NGC 1275 has such a peculiar shape, some astronomers previously had suspected that it actually might be two galaxies—a giant elliptical galaxy and a smaller spiral galaxy—passing through one another. In fact, elliptical galaxies in general may result from the merger of several spiral galaxies. Holtzman suggested that the clusters may have formed as a result of just such a merger, or collision. The fact that elliptical galaxies can contain a hundred times more globular clusters than spiral galaxies lends further support to the notion that galaxy collisions also create new globular clusters.

Also based upon images from the HST, astronomers asserted they'd found intriguing evidence that a black hole, weighing over 2.6 billion times the mass of the Sun, exists at the center of the giant elliptical galaxy M87 (so named because it was the 87th object in Charles Messier's 1774 catalogue of non-stellar objects). The images showed that stars become strongly concentrated toward the center of M87, as if drawn into the center and held there by the gravitational field of a massive black hole.

M87 is at the center of a nearby cluster of galaxies in the constellation of Virgo, 52 million light years distant, and contains more than 100 billion stars. One of the brightest galaxies in the local universe, M87 is visible in even small telescopes. Early in the twentieth century, astronomers discovered a gigantic plume, or 'jet,' of plasma apparently ejected out of the M87 nucleus. Later, the jet and nucleus were found to emit strong radio and X-ray radiation. However, the nature of the central 'engine' of this activity had long remained a mystery.

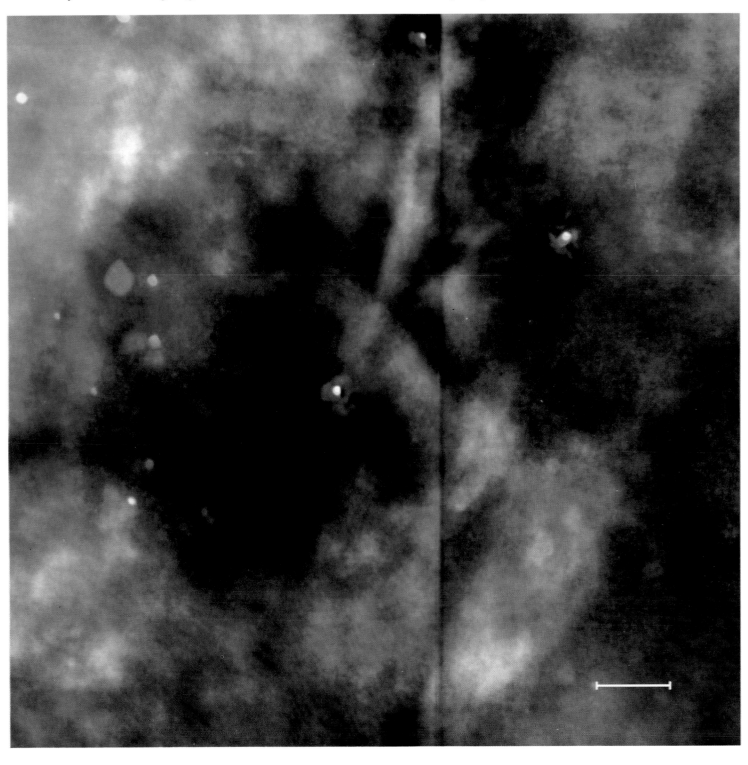

Stars, once freely orbiting in and out of the M87 core, are gradually being pulled towards the center and then into orbits closely bound to the black hole. As the core of the galaxy smoothly collapses inward, the density of stars near the center becomes extreme. Some of these stars eventually may be consumed by the black hole, further fueling its growth. This leads to the theory that one way to look for a black hole is to search for a strong concentration of starlight at the center of a galaxy.

In April 1992, evidence was found that a black hole, weighing three million times the mass of the Sun, existed at the center of nearby elliptical galaxy M32. Like M87, images of M32 showed that the stars in this galaxy became extremely concentrated toward the nucleus. This central structure resembled the gravitational 'signature' of a massive black hole. The presence of a black hole in an ordinary galaxy like M32 may mean that inactive black holes are common to the centers of galaxies. Astronomers considered M32 an interesting 'laboratory' for testing theories about the formation of massive black holes. Image analysis was conducted by Dr Tod Lauer of the National Optical Astronomy Observatory at Tucson, Dr **Sandra Faber** of the University of California at Santa Cruz and other members of the HST Wide Field/Planetary Camera (WFPC) Imaging Team.

M32 is quite small and compact, as elliptical galaxies go, containing about 400 million stars within a diameter of only 1000 light years. At a distance of 2.3 million light years, M32 is one of the closest neighbors to the Milky Way Galaxy.

M32 is a satellite of the great spiral galaxy in Andromeda (M31), which dominates the small group of galaxies of which the Milky Way is a member. M31 can be seen with the naked eye as a spindle-shaped 'cloud' the width of the full moon, and its small companion M32 can be seen with a small telescope. M32 has long been among the best candidates for a galaxy with a massive central black hole, a theory first proposed in 1987 by **Dr John Tonry** of MIT, and independently by **Dr Alan Dressler** of the Observatories of the Carnegie Institution at Washington, DC, and by **Dr Douglas Richstone** of the University of Michigan at Ann Arbor. Their observations, made with ground-based telescopes, showed an abrupt increase in the orbital velocities of stars towards the center of M32. This data led them to conclude that M32 must have a strong, but unseen, concentration of mass at its center. A black hole at least several million times the mass of the Sun was the most likely type of object matching such characteristics.

These ground-based images, however, did not have enough resolution to detect the effects of a massive black hole on the structure of M32 in clear detail. They found that the density of stars in the nucleus of M32 appeared to increase steadily towards the center, with no sign of leveling off. These results were very similar to the predictions for what a massive black hole would do to the central structure of a galaxy.

A black hole at the center of M32 would have the paradoxical effect of stabilizing the galaxy's nucleus, because the stars orbit so rapidly around the black hole, they move past each other too quickly to gravitationally capture each other or collide. The black hole thus keeps the center of a galaxy 'stirred up.' In the absence of a black hole, however, the stars would move slowly enough to attract each other gravitationally. Collisions between stars become much more frequent, and heavier, slower moving stars sink to the center of the galaxy, causing it to collapse. The fate of the collapsing core is uncertain. One possibility is that binary stars formed

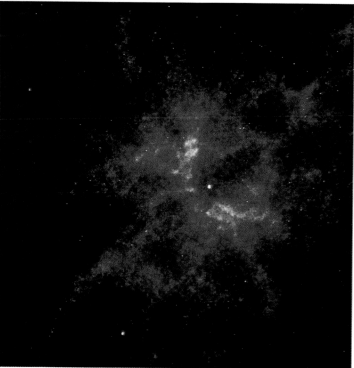

Opposite: **A HST image of material streaming away from a young star in the Orion Nebula.** *At top:* **The central core and jet of the giant elliptical galaxy M87.** *Above:* **One of hottest stars in the galaxy known as 'the NGC 2440 nucleus.**

during the collapse would provide enough kinetic energy to halt the collapse by transferring momentum to single stars. This would make the core rebound, like a rubber ball that has been squeezed and then relaxed.

An alternative possibility is that runaway merging of stars would occur during core collapse, leading to the formation of a black hole in any case. If so, this would rule out alternative explanations that don't require a black hole. If the core is really unstable, the researchers would expect to find evidence of merged and captured stars called 'blue stragglers.' The shape of the starlight distribution at the core would also be different from that which the HST detected.

The search for such supermassive black holes in the cores of galaxies is one of the primary missions of NASA's Hubble

Space Telescope. By investigating both active and quiescent galaxies, astronomers gain a better idea of the conditions and events which led to the formation and growth of supermassive black holes.

In May 1992, HST revealed a new class of objects in the universe: a grouping of gigantic star clusters produced by the collision of galaxies. Images of the core of the peculiar galaxy Arp 220 showed that stars were produced at a furious rate from the dust and gas supplied by the interaction of two galaxies. This discovery was made by **Dr Edward Shaya** and graduate student **Dan Dowling** at the University of Maryland and the Wide Field/Planetary Camera team.

Astronomers had never before seen a 'starburst galaxy' in such detail. The core of Arp 220 promised to be a unique laboratory for studying supernovas—which are the self-detonation of massive stars—because they should explode frequently in gigantic, young clusters. Over time, the core of this galaxy should resemble a string of firecrackers popping off. This would provide astronomers an unprecedented opportunity to study the late evolution of massive stars, as well as possibly improve techniques for measuring distances to galaxies, which use supernovae as 'standard candle' distance indicators.

In June 1992, HST provided astronomers with what may have been their first direct view of an immense ring of dust which fuels a massive black hole at the heart of the spiral galaxy M51, located 20 million light years away. These observations were reported by **Dr Holland Ford** of Johns Hopkins University and the Space Telescope Science Institute, Baltimore, and his co-investigators on HST's Faint Object Spectrograph, at the 180th meeting of the American Astronomical Society in Columbus, Ohio.

Ford announced that 'Pictures of M51, taken with HST's Planetary Camera, show a striking, dark X silhouetted across the nucleus. The X feature is due to absorption of light by dust

Below: A HST view of the core of galaxy NGC 1068. A schematic cone of radiation is superimposed over the hidden nucleus which is believed to contain a massive black hole. Opposite: A cluster of young stars in the 30 Doradus Nebula.

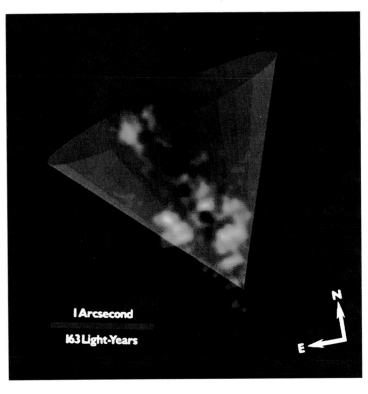

and marks the exact position of the nuclear black hole. If these ideas are correct, M51 provides the first direct view of a torus (ring of material) which both fuels a massive black hole and hides the hole from direct view from anyone outside the narrow cone of light emitted from the near vicinity of the black hole.'

Commonly called the Whirlpool Galaxy, M51 is one of the nearest and brightest galaxies, having an angular diameter one-third the width of the full Moon. The galaxy is spectacular because it is tilted nearly face-on to Earth, allowing for an unobstructed view of its bright core. M51 is especially noteworthy because its well-defined spiral arms are unusually bright, and the end of one of the spiral arms projects across a small, dusty and distorted satellite galaxy.

Previous observations, made with both radio and optical telescopes, have revealed energetic activity in the core of the galaxy. Hot, ionized gas in the center of M51 is moving at speeds of up to two million miles per hour. Ground-based observations also show a pair of radio and corresponding optical 'bubbles' that form a double-lobed structure across the nucleus.

The new HST images now offer the best glimpse yet of the near-vicinity of the 'powerhouse' driving these fireworks. The pictures reveal an hour-glass structure formed by two bright beacons of light that are so energetic they cause interstellar gas caught in their beams to glow through ionization. This double cone 'searchlight' is bisected by the widest bar of the dark X. Ford suggests that the dark band in the X, which is perpendicular to the ionization cone, may be 'a rotating torus (or ring) of cold gas and dust seen edge-on.'

The second bar of the X is both interesting and puzzling. The dust in this linear feature could be a second disk seen edge-on, or possibly rotating gas and dust in M51 interacting with the jets and ionization cones. 'The safest interpretation is that the slash is a caution sign, showing that we do not yet fully understand what is happening in the center of M51,' said Ford.

The edge-on torus, estimated to be 100 light years across, hides the black hole and its disk of infalling hot gas. This accretion disk, buried deep inside the torus, is presumably the source of the ionizing radiation. The dusty ring confines the radiation from the accretion disk so that it can escape only through the 'donut hole' of the torus as a pair of oppositely directed cones of light.

The ring also determines the axis of a jet of material being accelerated away from the black hole. Like a top tipped on its side, the dust ring is tilted so that it is perpendicular to the plane of M51. The high-speed jet thus lies in the galaxy's plane and plows into the gas and dust in the M51 disk. The jet inflates a bubble of hot gas on either side of the black hole. This is analogous to a fire hose directed against a large pile of sand. The fire hose inflates a cavity of water and sand, which expands and advances into the pile. The resulting optical and radio emission from this 'blowtorch' in M51 is several times brighter than the radio emission in the center of our own Milky Way galaxy. This means that the M51 black hole is more energetic than the million-solar-mass black hole suspected to lie at the center of the Milky Way galaxy.

By July 1992, a serendipitous survey of the heavens with the HST was uncovering remote and unusual galaxies never before resolved by optical telescopes on Earth. HST revealed an unusual variety of shape and structure in these distant galaxies, which previously appeared as fuzzy blobs in ground-based sky surveys. These early results could lead to a much

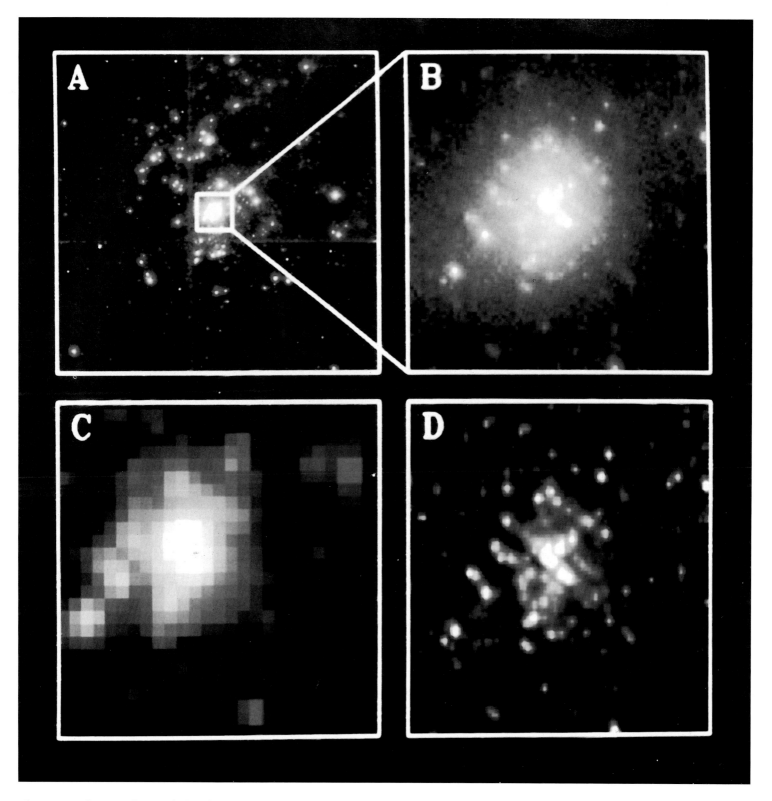

clearer understanding of the formation and evolution of galaxies.

Then, in November 1992, the Hubble observed what was the most distant known galaxy in the universe. It revealed a chain of luminous knots in the core of this galaxy—one that existed in the infancy of the universe and is located more than 10 billion light years from Earth. 'These knots could be giant clusters of stars. If that is so, then each knot would contain about 10 billion stars and would be about 1500 light years across,' said **Dr George Miley** of Leiden University in the Netherlands, the leader of the international astronomer team which examined the galaxy.

An alternative theory is that the knots are gas or dust clouds caught in a 'searchlight' beam of energy from a massive black hole hidden at the galaxy's core. The galaxy's great distance from earth indicates that it was formed only one or two billion years after the Big Bang, which marked the beginning of the observable universe. Most galaxies probably formed during this early epoch.

The galaxy, designated 4C 41.17, is also known as a radio galaxy. Radio galaxies produce powerful, extended radio emissions. Several have been discovered by this international team in the past few years at great distances from the Earth. In the case of 4C 41.17, astronomers presume that a massive black hole, rotating in the core of the galaxy, is producing twin jets of particles moving at enormous speed. The energy from the jets would be the source of the radio emissions.

These corresponding images suggest that the high velocity particle jets compress gas and dust along their paths, triggering the formation of new stars. This would account for the elongated optical appearance of the galaxy. If this explanation is accurate, the knots along the jet paths would be clusters of stars in 'enormous numbers, the products of the highly disturbed inner region of the primeval galaxy,' Miley said.

'It also is possible,' said Miley, 'that the light photographed by the HST is not due to stars along the jet paths, but rather is light from a disk of material surrounding the black hole which is being scattered off clouds of gas or dust. An active galactic nucleus of this description is called a quasar. It is hidden from optical view by a thick dust shroud which allows light to escape only along the radio or jet axis.'

The Hubble can help discriminate between these possibilities by further studying the colors and other properties of these and similar objects. After the scheduled Space Shuttle servicing mission for the Hubble in late 1993, HST then can be used to carry out detailed studies of many galaxies at distances comparable with 4C 41.17 'More than 50 are now known,' said Miley. 'Observing them with the renewed Hubble would provide us with an important new window through which we can glimpse the early history of our universe.'

In December 1992, the HST uncovered the strongest evidence yet that many stars may form **planetary systems**. **Dr C Robert O'Dell** of Rice University in Houston and colleagues discovered extended disks of dust around 15 newly-formed stars in the Orion Nebula starbirth region 1500 light years away. Such disks are a prerequisite for the formation of solar systems like Earth's.

'These images provide the best evidence for planetary systems,' O'Dell said. 'The disks are a missing link in our understanding of how planets like those in our Solar System form. Their discovery establishes that the basic material of planets exists around a large fraction of stars. It is likely that many of these stars will have planetary systems.'

According to current theories, the dust contained within the disks eventually agglomerates to make planets. Earth's Solar System is considered a relic of just such a disk of dust that accompanied the Sun's birth 4.6 billion years ago. Before the Hubble discovery, protoplanetary disks had been con-

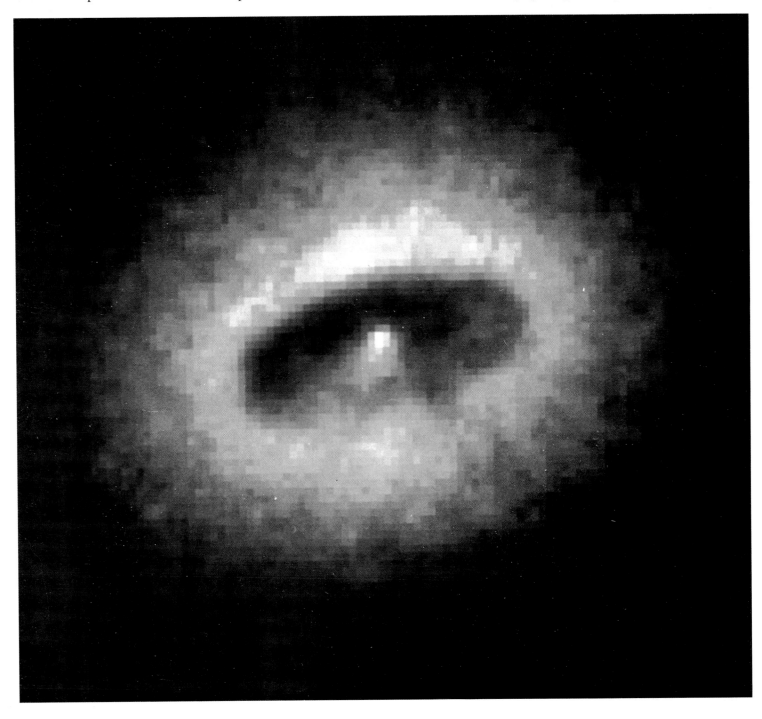

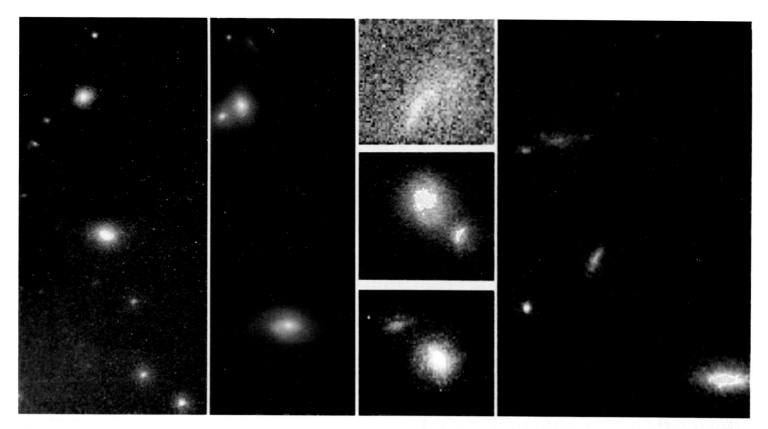

Left: **The HST captured this view of a suspected black hole at the core of galaxy NGC 4261 in the Virgo cluster.** *Above:* **Individual embryonic galaxies.** *Right:* **An FOC image of a distant quasar seen through the gravitational lens G2237+0305.**

firmed around only four stars: Beta Pictoris, Alpha Lyrae, Alpha Piscis Austrini and Epsilon Eridani. Unlike these previous observations, Hubble observed newly formed stars less than a million years old which are still contracting out of primordial gas. Hubble's images provide direct evidence that dust surrounding a newborn star has too much spin to be drawn into the collapsing star. Instead, the material spreads out into a broad, flattened disk. These young disks signified an entirely new class of object uncovered in the universe.

Hubble could see the disks because they are illuminated by the hottest stars in the Orion Nebula, and some of them are seen in silhouette against the bright nebula. However, some of these protoplanets are bright enough to have been seen previously as stars by ground-based optical and radio telescopes, but their true nature was not recognized until the Hubble discovery.

Each protoplanet appears as a thick disk with a hole in the middle where the cool star is located. Radiation from nearby hot stars 'boils off' material from the disk's surface at a rate equal to about one-half the mass of Earth per year. This material is then blown back into a comet-like tail by a stellar 'wind' of radiation and subatomic particles streaming from nearby hot stars.

Based on this erosion rate, O'Dell estimated that a protoplanet's initial mass would be at least 15 times that of the giant planet Jupiter.

On 7 January 1993, it was announced that astronomers using the Hubble believed that a galaxy they had observed for a decade actually was composed of *two* merged galaxies and that their collision provided new fuel for a massive black hole, which is spewing out a jet of gas and other matter 240,000 light years long. The galaxy, Markarian 315, is located about 500 million light years from Earth.

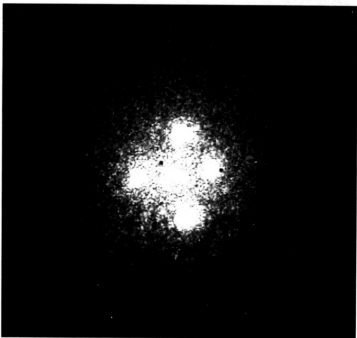

The collision and refueling theory emerged after the Hubble revealed that the galaxy had a double nucleus, or two core-like regions. The brighter core-like region is believed to harbor the massive black hole which accounts for the tremendous amounts of energy produced by the galaxy. The fainter nucleus is considered to be the surviving core of a galaxy that recently merged into Markarian 315.

'The galaxy's active core presumably harbors a black hole which has been refueled by the galactic collision,' said **Dr Jack MacKenty**, Assistant Scientist at the Space Telescope Science Institute. 'The Hubble images provide support for the theory that the jet-like feature may be a "tail" of gas that has been stretched out by tidal forces between the two galaxies as they have interacted,' explained MacKenty. 'The jet feature is most likely a remnant of a merger between Markarian 315 and a smaller galaxy.'

APPENDIX 1
THE PLANETS

Name	Diameter	Average Distance from the Sun	Number of Known Moons	Closest Visit by a Spacecraft from Earth
Mercury	3031 mi (4878 km)	36 million mi (58 million km)	0	Mariner 10 (1974)
Venus	7521 mi (12,104 km)	67 million mi (108 million km)	0	Venera 11-14 Landers (1978-82)
Earth	7926 mi (12,756 km)	93 million mi (150 million km)	1	—
Mars	4212 mi (6794 km)	141 million mi (228 million km)	2	Viking 1, 2 Landers (1976)
Jupiter	88,650 mi (142,984 km)	483 million mi (779 million km)	16	Voyager 1, 2 (1979)
Saturn	74,565 mi (120,000 km)	885 million mi (1.4 billion km)	21+	Voyager 1, 2 (1980-81)
Uranus	32,116 mi (51,800 km)	1.7 billion mi (2.9 billion km)	15	Voyager 2 (1986)
Neptune	30,775 mi (49,528 km)	2.8 billion mi (4.5 billion km)	8	Voyager 2 (1989)
Pluto	1375 mi (2200 km)	3.6 billion mi (5.9 billion km)	1	None (None planned)

APPENDIX 2
THE PLANETARY MOONS OF THE INNER SOLAR SYSTEM

Name	Discoverer/ Date of Discovery	Diameter	Distance from Planet
Mercury	None		
Venus	None		
Earth			
Luna (the Moon)	prehistoric	2160 mi (3476 km)	252,698 mi (406,676 km)
Mars			
Phobos	Asaph Hall, 1877	14 mi (23 km)	5760 mi (9270 km)
Deimos	Asaph Hall, 1877	7.5 mi (12 km)	14,540 mi (23,400 km)

APPENDIX 3
THE MOONS OF JUPITER

Name	Discoverer/Date of Discovery	Diameter	Distance from Jupiter
Metis	Project Voyager, 1979	30 mi (49 km)	79,750 mi (127,600 km)
Adrastea	Project Voyager, 1979	21 mi (35 km)	83,030 mi (134,000 km)
Amalthea	Edward Barnard, 1892	103 mi (166 km)	112,655 mi (181,300 km)
Thebe	Project Voyager, 1979	47 mi (75 km)	137,690 mi (222,000 km)
Io	Galileo Galilei, 1610	2257 mi (3632 km)	261,970 mi (421,600 km)
Europa	Galileo Galilei, 1610	1942 mi (3126 km)	416,877 mi (670,900 km)
Ganymede	Galileo Galilei, 1610	3278 mi (5276 km)	664,867 mi (1.1 million km)
Callisto	Galileo Galilei, 1610	2995 mi (4820 km)	1.2 million mi (1.9 million km)
Leda	Charles Kowal, 1974	5 mi (8 km)	6.9 million mi (11.1 million km)
Himalia	CD Perrine, 1904	105 mi (170 km)	7.1 million mi (11.5 million km)
Lysithea	SB Nicholson, 1938	12 mi (19 km)	7.3 million mi (11.7 million km)
Elara	CD Perrine, 1905	50 mi (80 km)	7.3 million mi (11.7 million km)
Ananke	SB Nicholson, 1951	11 mi (17 km)	12.8 million mi (20.7 million km)
Carme	SB Nicholson, 1938	15 mi (24 km)	13.9 million mi (22.4 million km)
Pasiphae	PJ Melotta, 1908	17 mi (27 km)	14.5 million mi (23.3 million km)
Sinope	PJ Melotta, 1914	13 mi (21 km)	14.7 million mi (23.7 million km)

APPENDIX 4
THE MOONS OF SATURN

Name	Discoverer/Date of Discovery	Diameter	Distance from Saturn
Atlas	Project Voyager, 1980	19 mi (30 km)	85,544 mi (137,670 km)
Prometheus	Project Voyager, 1980	137 mi (220 km)	86,589 mi (139,353 km)
Pandora	Project Voyager, 1980	56 mi (90 km)	88,048 mi (141,700 km)
Epimetheus	Project Voyager, 1980	40 mi (65 km)	94,089 mi (151,422 km)

(continued on page 182)

APPENDIX 4
THE MOONS OF SATURN (continued)

Name	Discoverer/Date of Discovery	Diameter	Distance from Saturn
Janus	Project Voyager, 1980	60 mi (95 km)	94,120 mi (151,472 km)
Mimas	William Herschel, 1789	242 mi (390 km)	115,326 mi (185,600 km)
Enceladus	William Herschel, 1789	311 mi (500 km)	147,948 mi (238,100 km)
Tethys	Giovanni Cassini, 1684	652 mi (1050 km)	182,714 mi (292,342 km)
Telesto	Project Voyager, 1980	9 mi (15 km)	183,118 mi (294,700 km)
Calypso	Project Voyager, 1980	9 mi (15 km)	217,480 mi (350,000 km)
Dione	Giovanni Cassini, 1684	696 mi (1120 km)	234,567 mi (377,500 km)
Helene	Project Voyager, 1980	20 mi (32 km)	234,915 mi (378,060 km)
Rhea	Giovanni Cassini, 1672	951 mi (1530 km)	327,586 mi (527,200 km)
Titan	Christiaan Huygens, 1655	3200 mi (5150 km)	759,067 mi (1.2 million km)
Hyperion	GP Bond and William Lassell, 1848	155 mi (250 km)	921,493 mi (1.5 million km)
Iapetus	Giovanni Cassini, 1671	905 mi (1460 km)	2.2 million mi (3.6 million km)
Phoebe	William Pickering, 1898	137 mi (220 km)	8 million mi (13 million km)

APPENDIX 5
THE MOONS OF URANUS

Name	Discoverer/Date of Discovery	Diameter	Distance from Uranus
Cordelia	Project Voyager, 1986	25 mi (40 km)	30,882 mi (49,700 km)
Ophelia	Project Voyager, 1986	31 mi (50 km)	33,429 mi (53,800 km)
Bianca	Project Voyager, 1986	31 mi (50 km)	36,785 mi (59,200 km)
Juliet	Project Voyager, 1986	37 mi (60 km)	38,400 mi (61,800 km)
Desdemona	Project Voyager, 1986	37 mi (60 km)	38,959 mi (62,700 km)
Rosalind	Project Voyager, 1986	50 mi (80 km)	40,140 mi (64,600 km)
Portia	Project Voyager, 1986	50 mi (80 km)	41,072 mi (66,100 km)

(continued on facing page)

APPENDIX 5
THE MOONS OF URANUS (continued)

Name	Discoverer/Date of Discovery	Diameter	Distance from Uranus
Cressida	Project Voyager, 1986	37 mi (60 km)	43,433 mi (69,900 km)
Belinda	Project Voyager, 1986	37 mi (60 km)	46,789 mi (75,300 km)
Puck	Project Voyager, 1985	106 mi (170 km)	53,437 mi (86,000 km)
Miranda	Gerard Kuiper, 1948	217 mi (150 km)	80,716 mi (128,282 km)
Ariel	William Lassell, 1851	721 mi (1160 km)	118,358 mi (190,900 km)
Umbriel	William Lassell, 1851	739 mi (1190 km)	165,284 mi (266,000 km)
Titania	William Herschel, 1787	998 mi (1610 km)	271,104 mi (436,300 km)
Oberon	William Herschel, 1787	961 mi (1550 km)	326,507 mi (583,400 km)

APPENDIX 6
THE MOONS OF NEPTUNE

NOTE: The numerically designated moons are listed in order of distance, but numbered in order of discovery.

Name	Discoverer/Date of Discovery	Diameter	Distance from Neptune
1989N6 (Naiad)	Project Voyager, 1989	30 mi (50 km)	14,400 mi (23,000 km)
1989N5 (Thalassa)	Project Voyager, 1989	60 mi (90 km)	15,500 mi (25,000 km)
1989N3 (Despina)	Project Voyager, 1989	85 mi (135 km)	16,980 mi (27,150 km)
1989N4 (Galatea)	Project Voyager, 1989	100 mi (160 km)	23,180 mi (37,100 km)
1989N2 (Larissa)	Project Voyager, 1989	125 mi (200 km)	30,080 mi (48,100 km)
1989N1 (Proteus)	Project Voyager, 1989	260 mi (420 km)	57,780 mi (92,500 km)
Triton	William Lassell, 1846	1690 mi (2700 km)	205,020 mi (329,880 km)
Nereid	Gerard Kuiper, 1949	300 mi (500 km)	3.4 million mi (5.5 million km)

APPENDIX 7
THE WORLD'S LARGEST REFLECTOR TELESCOPES

Telescope/Location	Aperture in Inches	(Meters)	Date of Completion
Bolshoi Teleskop Mount Pastuskhov, Zelenchukskaya, Russia	236	(6.00)	1976
George Ellery Hale Telescope Mount Palomar, California	200	(5.08)	1948
Multiple Mirror Telescope Whipple Observatory Mount Hopkins, Arizona	177	(4.49)	1979
William Herschel Telescope Observatorio Roque de los Muchachos La Palma, Canary Islands, Spain	165	(4.20)	1987
Inter-American Observatory Cerro Tololo, Chile	158	(4.00)	1976
Anglo-Australian Observatory Siding Spring, Australia	154	(3.90)	1974
Nicholas U Mayall Telescope Kitt Peak National Observatory, Arizona	148	(3.80)	1973
United Kingdom Infrared Telescope (UKIRT) Mauna Kea, Hawaii	148	(3.80)	1979
Canada-France-Hawaii Infrared Telescope Mauna Kea, Hawaii	142	(3.60)	1979
European Southern Observatory Cerro La Silla, Chile	142	(3.60)	1976
German-Spanish Astronomical Center Calar Alto, Spain	138	(3.50)	1985
C Donald Shane Telescope Lick Observatory Mount Hamilton, California	120	(3.05)	1959
NASA Infrared Telescope Mauna Kea, Hawaii	118	(3.00)	1979
McDonald Observatory Mt Locke, Texas	107	(2.70)	1968
Shajn Telescope Crimean Astrophysical Observatory Simeis, Ukraine	102	(2.60)	1961
Byurakan Astrophysical Observatory Mt Aragatz, Armenia	102	(2.60)	1976
John D Hooker Telescope Mount Wilson, California	100	(2.54)	1917
Irenee duPont Telescope Los Compañas Observatories Cerro Las Campanas, Chile	100	(2.54)	1976
Isaac Newton Telescope* Observatorio Roque de los Muchachos La Palma, Canary Islands, Spain	98	(2.49)	(1967) 1977*
Kitt Peak National Observatory Arizona	90	(2.29)	1969
University of Hawaii Telescope Mauna Kea, Hawaii	88	(2.24)	1970
McDonald Observatory, Texas	82	(2.08)	1939
Haute-Provence, France	74	(1.88)	1958
Mount Stromlo, Australia	74	(1.88)	1955

*Built in Herstmonceux, England but moved to Spain in 1977

APPENDIX 8
THE WORLD'S LARGEST REFRACTOR TELESCOPES

Telescope/Location	Aperture in Inches	(Meters)	Focal Length in Feet	(Meters)	Date of Completion
Yerkes Telescope, Williams Bay, Wisconsin	40	(1.02)	63	(19.4)	1897
Lick Observatory, Mount Hamilton, California	36	(0.92)	58	(17.9)	1888
Meudon Telescope, Paris, France	33	(0.85)	53	(16.3)	1896
Potsdam, Germany	32	(0.82)	39.2	(12.0)	1905
Allegheny Telescope, Pittsburgh, Pennsylvania	30	(0.80)	46	(14.2)	1914
Bischoffsheim Telescope, Nice, France	30	(0.80)	59	(18.1)	1887
Imperial Observatory, Pulkova, Russia	30	(0.80)	46	(14.2)	1886
Royal Observatory, Greenwich, England	28	(0.72)	28	(8.6)	1894
Bloemfontein, South Africa	27	(0.69)	40	(12.2)	1928
Vienna, Austria	27	(0.69)	34.2	(10.6)	1878
Johannesburg, South Africa	26.2	(0.67)	35	(10.8)	1926
Herstmonceux, England	26	(0.67)	22	(6.7)	1897
McCormick, United States	26	(0.67)	32.2	(9.9)	1873
Johannesburg, South Africa	26	(0.67)	36	(11.0)	1925

APPENDIX 9
THE WORLD'S LARGEST RADIO TELESCOPES

Telescope/Location	Aperture in Feet	(Meters)	Date of Completion
Very Large Array, Socorro, New Mexico	82x27	(24.6x27)	1979
Arecibo, Puerto Rico	1000	(300.0)	1963
Effelsberg, Germany	328	(98.4)	1976
Green Bank, West Virginia	300	(90.0)	1962
Jodrell Bank, England	250	(75.0)	1957
Parkes, New South Wales, Australia	210	(63.0)	1961
Green Bank, West Virginia	150	(45.0)	1966
Owens Valley, California	130	(39.0)	1968
Haystack, Massachusetts	120	(36.0)	1964
Vermilian River, Illinois	120	(36.0)	1971

APPENDIX 10
THE CONSTELLATIONS

Name	English Equivalent	Abbreviation	Area (Square Degrees)
Andromeda	Daughter of Cepheus	And or Andr	722
Antlia	The Air Pump	Ant or Antl	239
Apus	Bird of Paradise	Aps or Apus	206
Aquarius	Water Bearer	Aqr or Aqar	980
Aquila	The Eagle	Aqi or Aqil	652
Ara	The Altar	Ara or Arae	237
Aries	The Ram	Ari or Arie	441
Auriga	The Charioteer	Aur or Auri	657
Bootes	The Bear Driver	Boo or Boot	907
Caelum	The Sculptor's Chisel	Cae or Cael	125
Camelopardus	The Giraffe	Cam or Caml	757
Cancer	The Crab	Cnc or Canc	506
Canes Venatici	The Hunting Dogs	CVn or C Ven	465
Canis Major	The Greater Dog	CMa or C Maj	380
Canis Minor	The Lesser Dog	CMi or C Min	183
Capricornus	The Goat	Cap or Capr	414
Carina	The Keel (of Argo Navis)	Car or Cari	494
Cassiopeia	Mother of Andromeda	Cas or Cass	598
Centaurus	The Centaur	Cen or Cent	1060
Cetus	Sea Monster (The Whale)	Cet or Ceti	1281
Chamaeleon	The Chameleon	Cha or Cham	132
Circinus	The Compasses	Cir or Circ	93
Columba	The Dove	Col or Colm	270
Coma Berenices	Berenice's Hair	Com or Coma	386
Corona Australis	Southern Crown	CrA or CorA	128
Corona Borealis	Northern Crown	CrB or CorB	179
Corvus	The Crow or Raven	Crv or Corv	184
Crater	The Cup	Crt or Crat	282
Crux	Southern Cross	Cru or Cruc	68
Cygnus	The Swan	Cyg or Cygn	804
Delphinus	The Dolphin	Del or Diph	189
Dorado	The Swordfish	Dor or Dora	179
Draco	The Dragon	Dra or Drac	1083
Equuleus	The Foal	Equ or Equl	72
Eridanus	The River	Eri or Erid	1138
Fornax	The Laboratory Furnace	For or Forn	398
Gemini	The Twins	Gem or Gemi	514
Grus	The Crane	Gru or Grus	366
Hercules	Hercules	Her or Herc	1225
Horologium	The Clock	Hor or Horo	249
Hydra	The Water Serpent	Hya or Hyda	1303
Hydrus	The Water Snake	Hyi or Hydi	243
Indus	The American Indian	Ind or Indi	294
Lacerta	The Lizard	Lac or Lacr	201
Leo	The Lion	Leo or Leon	947
Leo Minor	The Lion Cub	LMi or LMin	232
Lepus	The Hare	Lep or Leps	290
Libra	The Scales of Balance	Lib or Libr	538
Lupus	The Wolf	Lup or Lupi	334
Lynx	The Lynx	Lyn or Lync	545
Lyra	The Lyre	Lyr or Lyra	286
Mensa	The Table Mountain	Men or Mens	153
Microscopium	The Microscope	Mic or Micr	210
Monoceros	The Unicorn	Mon or Mono	482

Musca	The Fly	Mus or Musc	138
Norma	The Carpenter's Square	Nor or Norm	165
Octans	The Octant	Oct or Octn	291
Ophiuchus	The Serpent Holder	Oph or Ophi	948
Orion	The Great Hunter	Ori or Orio	594
Pavo	The Peacock	Pav or Pavo	378
Pegasus	The Winged Horse	Peg or Pegs	1121
Perseus	The Hero, Son of Zeus	Per or Pers	615
Phoenix	The Phoenix	Phe or Phoe	469
Pictor	The Painter's Easel	Pic or Pict	247
Pisces	The Fishes	Psc or Pisc	889
Piscis Austrinus	The Southern Fish	PsA or PscA	245
Puppis	The Stern (of Argo Navis)	Pup or Pupp	673
Pyxis	The Compass Box	Pyx or Pyxi	221
Reticulum	The Net	Ret or Reti	114
Sagitta	The Arrow	Sge or Sgte	80
Sagittarius	The Archer	Sgr or Sgtr	867
Scorpius	The Scorpion	Sco or Scor	497
Sculptor	The Sculptor's Workshop	Scl or Scul	475
Scutum	The Shield	Sct or Scut	109
Serpens	The Serpent	Ser or Serp	637
Sextans	The Sextant	Sex or Sext	314
Taurus	The Bull	Tau or Taur	797
Telscopium	The Telescope	Tel or Tele	252
Triangulum	The Triangle	Tri or Tria	132
Triangulum Australe	The Southern Triangle	TrA or TrAu	110
Tucana	The Toucan	Tuc or Tucn	295
Ursa Major	The Greater Bear	UMa or UMaj	1280
Ursa Minor	The Lesser Bear	UMi or UMin	256
Vela	The Sail (of Argo Navis)	Vel or Velr	500
Virgo	The Virgin or Maiden	Vir or Virg	1294
Volans	The Flying Fish	Vol or Voln	141
Vulpecula	The Fox	Vul or Vulp	268

APPENDIX 11
SPECTRAL TYPES OF STARS

Type	Surface Temperature (Degrees C)	Color	Typical Star
W	36,000 plus	Greenish-white	Gamma Velorum
O	36,000 plus	Greenish-white	Zeta Puppis
B	28,000	Bluish	Spica
A	10,700	White	Sirius
F	7500	Yellowish	Beta Cassiopeiae
G (giant)	5200	Yellow	Epsilon Leonis
G (dwarf)	6000	Yellow	Our Sun
K (giant)	4230	Orange	Acturus
K (dwarf)	4910	Orange	Epsilon Eridani
M (giant)	3400	Orange-red	Betelgeuse
M (dwarf)	3400	Orange-red	Wolf 359
R	2300	Orange-red	U Cygni
N	2600	Red	S Cephei
S	2600	Red	R Andromedae

APPENDIX 12
THE BRIGHTEST STARS

NOTE: Stars are designated by constellation. Constellations are abbreviated.

Designation	Visual Magnitude	Spectral Classification	Distance in Light-Years	Proper Name (if any)
Sun	−26 73	G2	92,900,000 miles	Sun
Alpha CMa A	−1 45	A1	8.7	Sirius
Alpha Car	0 72	F0	98	Canopus
Alpha Boo	−0 06	K2	36	Arcturus
Alpha Lyr	0 00	A0	26.5	Vega
Alpha Cen A	0 01	G2	4.3	Alpha Centauri
Alpha Aur	0 06	B8	45	Capella
Beta Ori A	0 15	B8	900	Rigel
Alpha CMi A	0 35	F5	11.3	Procyon
Alpha Eri	0 51	B5	118	Achernar
Beta Cen AB	0 63	B1	490	Hadar
Alpha Tau A	0 86	K5	68	Aldebaran
Alpha Sco A	0 92v	M1	520	Antares
Alpha Vir	0 91v	B1	220	Spica
Beta Gem	1 13	K0	35	Pollux
Alpha PsA	1 16	A3	22.6	Fomalhaut
Alpha Cyg	1 25	A2	1600	Deneb
Beta Cru	1 28	B0	490	Beta Crucis
Alpha Leo A	1 35	B7	84	Regulus
Alpha Cru A	1 39	B1	370	Acrux
Alpha Cen B	1 40	dK1	4.3	Rigil Kentaurus
Epsilon CMa A	1 50	B2	680	Adhara
Lambda Sco	1 62	B1	310	Shaula
Gamma Ori	1 63	B2	470	Bellatrix
Beta Tau	1 66	B7	300	Elnath
Beta Car	1 67	A0	86	Miaplacidus
Epsilon Ori	1 70	B0	1600	Alnilam
Alpha Gru	1 73	B5	64	Al Na'ir
Zeta Ori AB	1 74	O9.5	1600	
Epsilon UMa	1 78	A0pv	68	Alioth
Alpha P	1 80	F5	570	Mirfak
Alpha UMa AB	1 80	K0	105	Dubhe
Delta CMa	1 80	F8	2100	
Gamma Vel A	1 82	WC7	520	
Epsilon Sgr	1 84	B9	124	Kaus Australis
Eta UMa	1 86	B3	210	Alkaid
Theta Sco	1 86	F0	650	
Alpha Ori	1 90	M2	520	Betelgeuse
Beta Aur	1 90	A2	88	
Gamma Gem	1 91	A0	105	
Alpha TrA	1 93	K2	82	Atria
Delta Vel AB	1 94	A0	76	
Alpha Pav	1 95	B3	310	
Alpha Hya	1 96	K4	94	Alphard
Alpha Gem A	1 97	A1	45	Castor
Epsilon Car	1 97	K0+B	340	Avior
Beta CMa	1 98	B1	750	
Gamma Leo AB	1 98	K0	90	

INDEX

Adams, John Couch 28, 29, 33, 39
Adrastea (moon of Jupiter) 129, 181
Air Density Explorer satellites *see* Explorer satellites 9, 24 and 39
Airy, GB 28, 29, 39
Aitken, RG 61
Aitoff equal area projection *68*
Albedos 42
Aldebaran (star) 27
Algol (binary star) 47, 61, 62, 64
Allegheny Observatory, USA 136, 185
Allen, Lew 169
Almagest 8, 13
Alouette satellites 88
Alpha 17
Alpha Centauri (star) 27, 58
Alpha Lyrae (star) 179
Alpha Orionis (star) 9
Alpha Piscis Austrini (star) 179
Altitude sextant *21*
Altazimuth 39
Amalthea (moon of Jupiter) 43, 124, 129, 181
Amata (Dione) 140
American Astronomical Society in Philadelphia 158, 172
American Astronomical Society of the United States 53, 116, 173, 176
American Physical Society 161
Amphitrite 29
Ananke (moon of Jupiter) 129, 181
Anaximander 11
Andromeda Galaxy (constellation) *see* M31 galaxy
Andromeda Nebula 60, 65, 69
Anglo-Australian Observatory, Australia 53, 54, 184
Anne's Spot (Saturn) 134
Annual parallax 24
Antarctica 33, *76*
Antares (star) 60
Antimatter 85, 92
Antoniadi, Eugenios 102
Aphrodite Terra (Venus) 111, 112, *113*
Apollo astronauts 31
Apollo Lunar Missions 31, 82, 116
  Apollo 17 mission 31
Appolonius of Perga 13
Aquarius (constellation) 29
Arab astronomers 8, 14, 62
Aratus 8
Arcetri, Italy 20
Arcturus (star) 22, 60
Arecibo Radio Telescope, Puerto Rico 57, 185
Areography (Martian cartography) 30
Argelander, WH 46, 58, 62
Argyre Planitia (Mars) *106*
Ariel (moon of Uranus) *106*, 183
Aries (constellation) 8, 13
Arietis (star) 17
A Ring *see* Saturn
A Ring Shepherd Moon *see* Atlas (moon of Saturn)
Aristarchus 13
Aristotle 12, 14, 19, 20
Aristyllus 12
Armillary sphere *21*
Arp 220 (galaxy) 176
Arsia Mons (Mars) *107*
Ascension Island 39
Ascraeus Mons (Mars) *107*
Association of Universities for Research in Astronomy (AURA) 168
  Space Telescope Science Institute 168, 173, 176, 179
Asteroid Belt 34, 98, 150
Asteroids 32, *34*, 34-35, 41, 42, 48, 49, 119
Astraea (asteroid) 32
Astrolabe 13, *18*
Astrology 6, 8
Astronomer Royal 22, 28
Astronomia Nova 17
Astronomical Optics 27
Astronomical Societies 53
Astronomical Society of the Pacific 53
Astrophysics 36-49
Atalanta Plain (Venus) 111
Athens Observatory, Greece 30
Atlas (moon of Saturn) 137, 181
AT Mic (star) 165
Atmosphere Explorers *see* Explorer satellites

Atomic particles 83, 85
AU Mic (star) 165
Auriga (constellation) 25, 65, 173
Auroras 22, 70, 74, 80, 83, 85, 92
Avicenna *14*
Azimuth quadrant *21*
Aztecs 14
Babylonian astronomers 6, 8, 9, 10, 12
Babylonian mythology 111
Bacon, Francis 15
Bacon, Roger 19
Bahcall, John 173
Bailey, S 64
Barnard, Edward Emerson 42, 43, 45, 60, 69, 129, 181
Barnard (galaxy) 5, *90*
Bath, England 24, 144
Bayer, Johann 8, 9, 62
Beer, Wilhelm 30, 41
Belinda (moon of Uranus) 183
Bellarmine, the Cardinal 20
Bennett, Charles 164
Bering Sea 39
Berlin Observatory, Germany 29, 34, 42, 45, 136
Bertsch, David 164, 165
Bessel, Friedrich William 27, 28, 40, 45, 58
Beta Lyrae (binary star) 62, 64
Beta Orionis (star) 9
Beta Persei *see* Algol
Beta Pictoris (star) 179
Beta Regio (Venus) 111
Betelgeuse (star) 36, 60, 64
Bianca (moon of Uranus) 182
Big Dome Telescopes *72-73*
Binary stars 26, 27, 60, 61, *61*, 62, 175
Bible, The 8
Bickerton, AW 69
Biela's Comet 32, 33, 45
Big Bang theory 158, 161, 164-165, 173, 177
Big Dipper (constellation) 60, 69
Binomial Theorem 21
Birmingham, J 65
Birt, WR 30
Bischoffsheim Telescope, France 185
Black holes 86, 89, 92, 161, 165, 166, 172, 174-179, *176*, *178*
BL Lacertae objects 161
Bloemfontein Telescope, South Africa 185
Bode, Johann Elert 32, 34
Bode's Law 32, 34
Boeing 116
Bolshoi Teleskop 54, 184
Bond, GP 61, 182
Bond, William Cranch 30, 39, 43, 136
Bonn Durchmusterung 46
Boston Observatory, USA 45
Brahe, Tycho 15, *16*, 17, *17*, 18, 21, 45
Bredechin, T 45
B Ring *see* Saturn
British Association for the Advancement of Science 30
British Astronomical Association
Brorsen's Comet 45
Bruce camera 39
Burnham, SW 61
Byurakan Astrophysical Observatory, Armenia 54, 184
Callipus 12
Callisto (moon of Jupiter) 18, 98, 119, 132, *132*, 133, *133*, 143, 181
Calypso (moon of Saturn) 182
Camac, Brittany 9
Cambridge Telescope, England 20, 28, 29, 36, 38
Camelopardalis (constellation) 161
Campbell, WW 41, 60, 62
Canada-France-Hawaii Telescope, USA 54, 184
Canali (canals) *see* Mars
Canes Venatici (constellation) 28
Canopus (star) 119
Cape Canaveral, USA 103, 116
Cape of Good Hope Observatory, South Africa 32, 40, 45, 46, 73, 74
Carme (moon of Jupiter) 129, 181
Carpenter, J 30
Carrington, RC 74
Cassini Division 20, 39, 42, 136
Cassini, Giovanni Domenico *see* Cassini, Jean Dominique

Cassini Interplanetary Explorer 150
Cassini, Jacques 20, 39
Cassini, Jean Dominique 20, 26, 42, 43, 102, 135, 136, 140, 182
Cassini spacecraft *156*
Cassiopeia (constellation) 15, 46
Castor (star) 36, 38
Celestial constants 101
Celestial equator 8, 12, 58
Celestial latitude 13
Celestial longitude 13
Celestial Police, Germany 32
Centaur launch vehicles 103, 104
Centaurus (constellation) 22
Ceres (asteroid) 32, 34, *34*, 35, 48
Cetus (constellation) 62
Chaldean astronomers 6, 8, 102
Challis, JC 28, 29
Chandler, SC 45
Charon (moon of Pluto) 49, 169
Chi Cephei (variable star) 64
Chi Cygni (binary star) 62, 64
Chinese astronomers 8, 9, 10, 23, 64, 65, 70
Chladni, EFF 32
Christian, Carol A *62*
Christy, James 49
Circinus (constellation) 92
Clark, AG 61
Classification of stars 60
Classification of variables 64
Clepsydra (water clock) 10, 17
Coal Sack 69
Coggia's Comet 45
Comet Bennett 94
Comet Giacobini-Zinner 33, 80
Comet Kohoutek 33
Comet of 1887 45
Comet Rendezvous/Asteroid Flyby (CRAF) 150, *157*
Comets 6, 8, 10, 17, 22, 25, 26, 28, 32, 33, 38, 45, 46, 64, 89, 90, 94, 136, 168
Comet Tago-Sato-Koska 89, 94
Common, AA 38, 60, 69
Comstock, Professor 31
Compton, Arthur Holly 159
Arthur Holly Compton Gamma Ray Observatory 159, *159*, 161, 164
Constellations 6, 8, 9, 186-187
Co-orbitals 136-137, 138
Copernicus, Nicholas 14-15, *16*, 17, *17*, 19, 20
Cordelia (moon of Uranus) 147, 182
Cordoba Observatory 61
Cornell University, USA 57
Corona *see* Sun
Coronal transit 80
Corona Borealis (constellation) 65
Coronagraph 47, *73*
Cosmic Background Explorer (COBE) 158, *162-164*
Cosmic background radiation 161, 164, 165
Cosmic dust 89, 90, 101, 158, 176, 177, 178, 179
Cosmic gamma ray bursts 161
Cosmic jets 172
Cosmic rays 80, 83, 92, 159
Cosmotheoros 20
Crab Nebula *93*, 161, 164, 165
Cressida (moon of Uranus) 183
Crimean Astrophysical Observatory, Ukraine 184
C Ring *see* Saturn
Crommelin, ACD 9
Cumana, Venezuela 33
Cuzzi, Jeff 138
Cygnus (constellation) 65, 69, 89, 92
Cygnus X-1 (binary star) 89
Daguerre, Louis 38
Dampier, Sir WC 8
Dark Spot 2 *see* Lesser Dark Spot (Neptune)
D'Arrest, Heinrich Ludwig 29, 148
Darwin, Charles 30
Darwin, GH 30
da Vinci, Leonardo 14
Davis, Don 143
Dawes, WH 29, 41, 60, 73
Deferent 13, 17
Deimos (moon of Mars) 41, 94, 180
de La Caille, Nicolas Louis 9
Delaunay, C 39
Delta Ring *see* Uranus
Delta Cephei (binary star) 62, 64

Dembowski, Baron E 61
Denning, WF 40, 42, 46
de Roberval, GP 39
Descriptive Astronomy 26
Deslandres, H 73
Desdemona (moon of Uranus) 182
Irenée duPont Telescope, Chile 54, 184
de Vico, F 40
Dialogue on the Two Chief Systems of the World, the Ptolemaic and the Copernican 20
Diana Chasma (Venus) 111
Differential Microwave Radiometers (DMR) 161, *162*
Diffraction grating 36
Diffuse Infrared Background Experiment (DIRBE) 158, *163*
Digges, Leonard 19
Dione (moon of Saturn) *138*, 140, *141*, 142, 182
Dollond, John 19
Donati, G 45, 58
Donati's Comet 45
Doppler, Christian Johann 38
Doppler's principle 47, 60
Dorpat Observatory 40
Double stars *see* Binary stars
Douglas, AE 41, 70
Dowling, Dan 176
Downing, AM 46
Draper, H 45, 60, 69
Draper Catalogue of Stellar Spectra 60, 61
Dressler, Alan 175
Dreyer, JL 69
Duncan, Douglas 173
Dynamics Explorer 88, *88*
Eagle Nebula *63*
Earl of Rosse 28, 31
Earth 9, 12, 13, 14, 17, 18, 20, 21, 23, 24, 25, 27, 30, 31, 32, 33, 38, 39, 40, 41, 45, 46, 73, 74, 79, 80, 82-88, 90, 91, 92, 94, 96, 102, 104, 106, 111, 112, 119, 121, 122, 124, 129, 131, 133, 134, 135, 136, 138, 140, 142, 143, 147, 148, 149, 150, 152, 156, 161, 165, 166, 169, 172, 173, 176, 177, 178, 179, 180
  Atmosphere 79, 80, 83, 88, 89, 92, 96, 97, 106, 111, 166, 169, 173
  Exosphere 96
  Ionosphere 79, 83, 86, 88, 92
  Magnetic field 74, 79, 83, 85, 86, 88, 92, 94, 96, 97, 98, 159
  Magnetotail 79, 80
  Mesosphere 79
  Polar wind 88, 98
  Stratosphere 79
  Thermosphere 79, 88
Earthshine 14, 41
Easton, C 69
Eccentrics Theory 13, 14, 15
Eclipse 6, 9, 13, 15, 18, 20, 38, 45, 47, 62, 64, 70, *71*, 76, *77*
Ecliptic 8, 9, 10, 11, 12, 13, 20, 25, 28, 30
Edinburgh, UK 19, 65
Effelsberg Radio Telescope, Germany 185
Egyptian astronomers 6, 9, 10, 11, 13, *16*
Einstein Observatory *see* High Energy Astronomy Observatories (HEAO 2)
Elara (moon of Jupiter) 129, 181
Electromagnetic radiations 82, 83, 89
Electromagnetic spectrum 165, 166
Elger, T Gwyn 30-31
Ellipse 18, 21, 33, 45
Enceladus (moon of Saturn) 136, 138, *139*, 140, *140*, 182
Encke, Johann Franz 22, 23, 32, 136
Encke's Comet 33
Encke Division 136, 138
Energetic Gamma Ray Experiment Telescope (EGRET) *158*, 161, 164
Epicycles 13, 14, 15, 17
Epimetheus (moon of Saturn) 137, 181
Epsilon Eridani (star) 179
Epsilon Ring *see* Uranus
Equatorial Telescope 40, *42*
Equinoxes 9, 12, 13, 17
Eratosthenes 13
Eridanus (constellation) 161
E Ring *see* Saturn
Eros (asteroid) 42
Eta Aquilae (variable star) 64
Eudoxus of Cnidus 8, 12, 13
Europa (moon of Jupiter) 18, 98, 119, *129*, *131*, 132, *132*, 150, 181

European astronomers 6, 10, 11, 14
European Celestial Observation Satellite 161
European Southern Observatory, Chile 54, 184
European Space Agency 79, 86, 169, 172
　　Faint Object Camera (FOC) 166, 169, 172, *179*
Evans, David 156
Explorer satellites *82-83*, 83-89, 92
　　Explorer 1 satellite 83
　　Explorer 6 satellite 83
　　Explorer 7 satellite 83
　　Explorer 8 satellite 86
　　Explorer 9 satellite 86
　　Explorer 10 satellite 83
　　Explorer 11 satellite 83, 85
　　Explorer 12 satellite 83
　　Explorer 14 satellite 83
　　Explorer 15 satellite 83
　　Explorer 17 satellite 86, *86*
　　Explorer 18 satellite *87*, 94
　　Explorer 20 satellite 86, *87*
　　Explorer 22 satellite 88
　　Explorer 24 satellite 86
　　Explorer 25 satellite 86
　　Explorer 26 satellite 83
　　Explorer 30 satellite 83
　　Explorer 31 satellite 86
　　Explorer 32 satellite 86
　　Explorer 37 satellite 83
　　Explorer 38 satellite *84*, 85
　　Explorer 39 satellite 86
　　Explorer 40 satellite 86
　　Explorer 42 satellite 85
　　Explorer 45 satellite 85
　　Explorer 47 satellite *89*
　　Explorer 48 satellite 85, *85*
　　Explorer 49 satellite 85
　　Explorer 51 satellite 88, *88*
　　Explorer 53 satellite 85
　　Explorer 54 satellite 88
　　Explorer 55 satellite 88
Extra-solar planets 166
Extraterrestrial life 101, 104
Extreme Ultraviolet Explorer Spacecraft (EUVE) 165, *165*
Extreme Ultraviolet spectrum 165
Faber, Sandra 175
Fabricius, D 62
Faculae 74
Faint Object Camera (FOC) *see* European Space Agency
Faint Object Spectrograph 176
Far Infrared Absolute Spectrophotometer (FIRAS) 158
Faye, HAE 79
Ferrel, W 39
Fichtel, Carl 161
Fine Guidance Sensors (FGS) 166
Fisk, Lennard A 169
Flora (asteroid) 32, 39, 40
Fluxions 21
Forbush 80
Ford, Holland 176
4C38.41 (quasar) 161
4C 41.17 (galaxy) 177-178
Foucault, Jean Bernard 19, 36, 38
Fraunhofer, Joseph 27, 36, 58, 79
F Ring *see* Saturn
F Ring Shepherd Moons *see* Prometheus and Pandora
Fuh Hsi 10
Fundamenta Astronomiae 27, 46
Fusion 80
Galactic center *90*
Galaxies 18, 25, 26, 28, 60, 69, 85, 86, 90, 101, 158, 161, 164, 165, 166, 168, 172-179
　　Disc 69
　　Double-armed spiral 69
　　Elliptical 86, 174, 175
　　Ring 69
　　Satellite 176
　　Spiral 28, 60, 69, 174, 175, 176
　　Twisted stream 69
Galileans *see* Jupiter
Galilei, Galileo 18-20, *20*, 24, 27, 124, 129, 134-135, 148, 150, 181
Galilei, Vincenzo 18
Galileo Regio (Ganymede) 133, *133*
Galileo spacecraft 150-156, *150-155*
Galle, Johann Gottfried 29, 39, 148
Gamma Aquilae (star) 69
Gamma Arietis (binary star) 60
Gamma Ray Observatory (GRO) *see* Arthur Holly Compton Gamma Ray Observatory

Gamma rays 80, 83, 85, 89, 92, 159, 161, 164, 165
Ganymede (moon of Jupiter) 18, 98, 116, 119, 129, 132-133, *132*, *133*, 138, 142, 143, 181
Gas Exchange Experiment 104
Gaspra (asteroid) 150, *154*
Gauss, Karl Friedrich 26-28, 34
Gautier, A 74
Geminga (neutron star) 164
Gemini (constellation) 25, *49*
Geminid meteor showers 46
Geomagnetosphere *see* Earth
Geomagnetotail *see* Earth
George III 25
German-American Roentgen Satellite 164, 165
German-Spanish Astronomical Center, Spain 54, 184
Gill, Sir David 39, 40, 46
Globular star clusters 22, 65, 92, 173-174
Gnomon 10, 11
Goddard High Resolution Spectrograph 172
Goddard Space Flight Center *see* National Aeronautics & Space Administration (NASA)
Goddard Ultraviolet Imaging Telescope (UIT) 160
Goodricke, J 61, 62
Gore, JE 62, 69
Gould, BA 58
Gould, FA 61
Grand Duke of Tuscany 134
Grand Canyon (Earth) 140
Grand Tour 116 *see also* Voyager Project
Gravitational Astronomy 28-29, 39
Gravitation lens G2237+0305 *179*
Great Andromeda Nebula *see* Andromeda Nebula
Great Caracole Observatory, Chichen Itza 15
Great Comet of 1882 45
Great Dark Spot (Neptune) 121, 148-149, *148*
Great Pyramid 9, 11, *11*
Great Nebula 63
Great Red Spot (Jupiter) 42, 43, 98, 101, 121, 124, *124*, 126, *128*, 134, 149
Greek astronomers 6, 8, 9, 10, 13, 22, 102
Greek mythology 6, 29, 102, 122, 129, 132, 137, 138, 140, 142, 144
Green Bank Radio Telescope, USA 57, 185
Greenhouse effect 108, 119
Greenwich Observatory *see* Royal Observatory, England
G Ring *see* Saturn
Hadley, John 19
Hague, The 20
Hale, George Ellery 47, 53, 57, 73, 74
George Ellery Hale Telescope *52*, 53, 54, 184
Hall, Asaph 41, *41*, 43, 180
Hall, Chester Moore 19
Halley, Sir Edmund 22, 23, 32
Halley's Comet *3*, 6, 8, 10, 22, *22*, 28, 32, 33, 46, 80
Halpern, Jules 164, 165
Hamal (star) 17
Harmonic Law 18
Harriott, Thomas 19
Harvard College Observatory, USA *37*, 38, 47, 58, 60, 61, 64, 136
　　Observatory Station, Peru 46
　　Observatory Station, South Africa 39
Harvard University Observatory, USA 30
Haute-Provence Telescope, France 184
Haystack Radio Telescope, USA 185
Heap, Sally 172, 173
Hebe (asteroid) 32
Heidelberg Observatory, Austria 42
Heis, E 46
Helene (moon of Saturn) 182
Heliometer 27, *44*, 45, 58
Heliopause *see* Sun
Helioseismologists *75*
Heliosphere *see* Sun
Helium stars 60
Hencke, Karl Ludwig 32
Henderson, T 27, 58
Henry brothers 60, 61
Hercules (constellation) 22, 92, 161
Herschel (Mimas) 138
Herschel, Caroline Lucretia 26, 58, 69
Herschel, Friedrich Wilhelm *see* Herschel, William
Herschel, Sir John 28, 30, 32, 33, 36, 58, 69, 73
Herschel, William 19, 24-26, *25*, 28, 29, 40, 41, 43, 58, 60, 62, 69, 70, 73, 102, 104, 138, 144, 182, 183

William Herschel Telescope 54, 184
Herstmonceux Telescope, England 185
Hesiod 8
Hevel, Johannes 9
Hevelius 25, 26, 45, 62
H Geminorum (star) 24, 25
High Energy Astronomy Observatories (HEAO) 80, 92
　　HEAO 1 satellite 92
　　HEAO 2 satellite 92, *93*
　　HEAO 3 satellite 80, *80*, 92
Himalia (moon of Jupiter) 129, 181
Hind, JR 65
Hindu astronomers 11
Hipparchus 8, 10, 11, 12, 13, 45
His Majesty's Astronomer 40
HK Tauri (star) 91
Holberg, Jay 122
Holmes, E 45
Holt, Stephen 164, 165
Holtzman, Jon 173
Homer 8, 108
Hooke, Robert 42, 60, 124
Hooker, John D 53
John D Hooker Telescope 53, 184
Johns Hopkins University, USA 168, 176
Horoscopic astrology 6
Horrebow, Christian 70
Howlett, F 74
Hubble, Edwin P 166
Edwin P Hubble Space Telescope (HST) 57, 166, *166*, *167*, 168, *168*, 169, *169*, *170-171*, 172-179, *179*
Huggins, Sir W 31, 41, 43, 45, 47, 58, 60, 65, 69, 76
Hughes Danbury Optical Systems 169
Huntress, Wesley 156
Hussey, Reverend TJ 28
Hussey, WT 61
Huygens, Christiaan 20, 41, 43, 60, 102, 135, 142, 182
Hyperion (moon of Saturn) 136, 182
Iapetus (moon of Saturn) 20, 135, 136, *141*, 182
Ida (asteroid) 150
Il Saggiatore (The Assayer) 20
Imperial Mathematician 17
Imperial Observatory, Russia 185
Index Catalogue 69
Indian astronomers 6
Inertial Upper Stage (IUS) rocket *111*
Inferior planets 9
Infrared 89, 90, 91, 94, 158, 165
Infrared Astronomy Satellites (IRAS) 90, *90*
Injun Explorer *see* Explorer 25 satellite
Innes, RTA 61
Inquisition, The Spanish 20
Institute for Advanced Study 173
Inter-American Observatory, Chile *3*, 54, *54*, 184
International Astronomical Union 32, 161
International Magnetospheric Study 79
International Sun-Earth Explorers (ISEE) 79
　　ISEE 1 spacecraft 79
　　ISEE 2 spacecraft 79
　　ISEE 3 spacecraft 79, 80, *80*
International Ultraviolet Explorer satellite (IUE) 86, 172
Interplanetary Monitoring Platforms (IMP Explorers) 85-86
　　IMP Explorer 18 85
　　IMP Explorer 33 86
　　IMP Explorer 35 86
　　IMP Explorer 47 *89*
Interplanetary spacecraft 150-157
Interstellar dust clouds 90, 91, 158, 176, 177, 178, 179
Interstellar gas 158, 159, 176, 177, 178, 179
Inverse Fluxions 21
Io (moon of Jupiter) 18, 98, 116, 119, 122, 129-131, *129*, *130*, *132*, 150, *154*, 181
Iris (asteroid) 32, 39, 40
Ishtar Terra (Venus) 111, *113*
Ithaca Chasma (Tethys) 140
Izanagi (Rhea) 142
Jansky, Karl 57, *57*
Janson, W 65
Janus (moon of Saturn) 137, 182
Japanese astronomers 64
Jeffreys, H 39
Jodrell Bank Radio Telescope, England 57, 185
Johannesburg Telescope, South Africa 185
Joint Institute for Laboratory Astrophysics (JILA) 173

Judicial astrology 6
Julianus 17
Juliet (moon of Uranus) 182
Juno (asteroid) 39, 40
Jupiter 8, 9, 11, *14*, 18, 20, 22, 24, 25, 32, 33, 34, 39, 42, 43, *43*, 45, 46, 80, 85, 89, 98, 101, *101*, 116-121, 122, 124-127, *124-129*, 134, 138, 143, 144, 147, 148, 150-156, *155*, *173*, 179, 180
　　Moons of 18, 20, 42, 43, 98, 101, 104, 116, 119, 126-133, *129*, 138, 143, 150, *154*, 181
　　Rings of 124, *127*
Jupiter-Saturn-Pluto mission 116
Jupiter-Uranus-Neptune mission 116
Kaiser, F 41
WM Keck Observatory, USA 57
Keeler, James E 39, 41, 43, 60, 69, 136
Keeler Gap 136
Kennedy Space Center 103
Kepler, Johannes 17-18, *18*, 19, 21, 34
Kepler's Three Laws 17, 18, 19, 21, 23, 39
Kepteyn, JC 46, 65
Kirchhoff, GA 36, 76
Kirkwood, Daniel 39, 42
Kitt Peak National Observatory, USA *2*, *7*, 54, 62, 78, 79, 160, 184
Klein, HJ 30
Koppernigk, Niklas *see* Copernicus, Nicholas
Kowal, Charles 129, 181
Kuiper, Gerrard Peter 45, 48, 122, 183
Labeled Release Experiment 104
Lagoon Nebula 66-67
Lagrangian satellites 136-137
Lakshmi Planvin (Venus) 111
Lalande, Joseph Jerome 29
Lamont, J 70, 73
Langley, Samuel Pierpont 31
Laplace, Pierre S 32, 39
Large Magellanic Cloud (galaxy) *64*, 90, 173
Lasers 88
Lassell, William 25, 29, *29*, 43, 45, 136, 182, 183
Lauer, Tod 172, 174
Law of Area 18
Law of Force 19
Law of Gravity 18, 21
Law of Motion 19
Least Squares 28
Leda (moon of Jupiter) 129, 181
Leiden University, Netherlands 177
Leo (constellation) 33, *61*
Leonid meteors 33, 46
Lesser Dark Spot (Neptune) 121, *148*, 149
Le Verrier, Urban Jean Joseph 28, 33, 46, 73
James Lick Observatory, USA 30, 38, 41, 43, *43*, 45, *50*, 50, 60, 61, 65, 69, 129, 184, 185
Lilienthel, Germany 26
Linne (Moon) 30
Linsky, Jeffrey 173
Lippershey, Hans 19
Little bang theories 165
Lockyer, J Norman 41, 60, 76
Lohrmann, Wilhelm Gottfried 30
Loki (Io) 131
Los Companas Observatories, Chile 184
Lowell, Percival 40, 41, 42, 43, 48, 97, 102, 132
Lowell Observatory, USA 41, 48, 50, 61, 173
Lunar Committee (England) 30
Lyman, CS 40
Lyot, B 47
Lyrid meteor showers 46
Lysithea (moon of Jupiter) 129, 181
M8 66-67
M20 *1*, 58
M31 (Andromeda Galaxy) 9, 33, *65*, 175
M32 (galaxy) 175
M 42 63
M45 *59*, *189*
M51 (Whirlpool Galaxy) 28, 176
M80 (globular cluster) 65
M81 (spiral galaxy) *160*
M87 (galaxy) 174-175, *175*
M99 (spiral nebula) 28
McDonald Observatory, USA 53, 184
McMath-Pierce Solar Telescope *77*, *78*, 79
MacKenty, Jack 179
Magellan spacecraft 109-115, *110-113*
Magnetograph *73*
Mahoney, Michael J 90
Marianas Trench (Earth) 111
Mariner-Jupiter-Saturn mission 116
Mariner-Jupiter-Uranus mission 116
Mariner spacecraft 94-98, 116, 119, 150, *157*
　　Mariner 2 spacecraft 94, *94*, 108
　　Mariner 4 spacecraft 94, 102, 156

INDEX

Mariner 5 spacecraft 94, *95*
Mariner 6 spacecraft 94, 102
Mariner 7 spacecraft 94, 102
Mariner 8 spacecraft *95*
Mariner 9 spacecraft 94-96, *97*, 102
Mariner 10 spacecraft *94*, 96-97, *96-98*, 180
Mariner 11 spacecraft 116
Mariner 12 spacecraft 116
Mariner Mark II *157*
Markarian 315 (galaxy) 179
Mars 8, 11, *14*, 24, 25, 30, 32, 34, 36, 39, 40, *40*, 41, 80, 89, 94, *97*, 102-108, *102*, *103*, *106*, *107*, 116, 142, 150, 156, *173*, 180
    Canals of *40*, 41, 94, 97, 102, 106, 132
    Moons of 41, 94, 180
Marsh, Kenneth 90
Mars Observer spacecraft 150, 156, *157*
Marseilles Telescope, France 38
Martin Marietta 116
Mastlin 18
Mathematical Astronomy 26
Mathematics 11, 12
Mathematike Syntaxis see Almagest
Mather, John C 158, 165
Ma Tuan Lin 11
Maxwell, James Clerk 39
Maxwell Mountains (Venus) 111, *112*
Nicholas U Mayall Telescope *2*, 54, *95*, 184
Mayer, C 60
Mayer, Simon 19
Mayer, Tobias 26
Megaliths 9
Melbourne Telescope, Australia 38, 46
Melnick 42 (star) 172, *173*
Melotta, PJ 129, 181
Mercury 8, 10, 11, 12, *14*, 25, 32, 40, 49, 70, 96-98, *97*, 102, 108, 119, 180
Meridian circle 27
Meridian photometer 58
Mesopotamia 11
Messier, Charles 9, 45, 174
Messier objects 9
Meteoric Astronomy 32-33
Meteoric Hpothesis 60
Meteors 18, 31, 32-33, *33*, 41, 46, 60, 136
Metis (moon of Jupiter) 129, 181
Meudon Observatory, France 74, 185
Meudon Telescope 47, 74, 185
Mexico 6
Micrometers 27, *27*
Micrometeoroids 83, 86
Microwave radiation 158, 161, 164, 165
Mihalov, John 101
Milan Observatory, Italy 40, 41
Miley, George 177-178
Milky Way 18, 25, 26, 60, 69, *69*, 86, 90, *90*, 92, 94, 158, 161, *161*, 164, 165, 173, 174, 175, 176
Mimas (moon of Saturn) 39, 42, 138, *139*, 140, 182
Mira Ceti (binary star) 24, 62, 64
Miranda (moon of Uranus) 138, *144*, *146*, 183
MIT 175
MK 421 (BL Lacertae object) 161
Monck, WHS 65
Der Mond: oder allgemeine vergleichende Selenographie 30
Montanari, G 62
Moon 6, 8, 9, 11, 12, 13, 14, 15, 17, 18, 19, 21, 22, 24, 26, 28, 30, 31, *31*, 32, 36, 38, 41, 47, 70, 73, 78, 82-83, 85, 89, 108, 112, 116, 119, 129, 136, 175, 176, 180
The Moon 30
Mounder, EW 70, 74
Mount Etna (Earth) 131, 138
Mount Everest (Earth) 111
Mount Pastukhov, Russia 54, 184
Mount St Helens (Earth) 131
Mount Stromlo Observatory, Australia 184
Mount Wilson Observatory, USA 50, 53, 54, 129, 184
Muller, Johannes 14, 42
Multiple Mirror Telescope 54, *54*, 184
Nasmyth, J 30, 74
National Aeronautics & Space Administration (NASA) 57, 79, 80, 82-157, 159, 161, 164, 165, 166, 168, 169, 172
    Advanced X-Ray Astrophysics Facility 169
    Ames Research Center 137, 138
    Goddard Space Flight Center 159, 164, 165, 168, 172
    Infrared Telescope, USA 54, 184
    Jet Propulsion Laboratory (JPL) 116-149, 156, 169

Marshall Space Flight Center 168
Office of Space Science and Application 168, 169
Search for Extra-Terrestrial Intelligence (SETI) 57
Space Environment Simulator *92*
Space Program *see* individual listings of spacecraft
Solar System Exploration Division 156
Ultraviolet and Visible Astrophysics Branch 165
National Astronomy & Ionosphere Center 57
National Institute of Standards and Technology (NIST) 173
National New Technology Telescope 54
National Optical Astronomy Observatories 54, 175
National Science Foundation 57
Nature magazine 90, 164
Nebulae 18, 22, 24, 25, 28, *28*, 29, 38, 47, 60, *63*, 65, 66-67, 69, 89, *91*, *93*, 161, 164, 169
    Spiral 28, *28*, 60, 69
Neison, E 30
Neptune 28-29, 32, 33, 34, 39, 43, 45, 48, 49, 98, 101, 116-122, 129, 134, 136, 144, 148-149, *148*, *149*, 180
    Moons of 29, 45, 49, 121, 122, 136, 183
    Rings of 121
Nereid (moon of Neptune) 45, 122, 183
The Netherlands 19, 90
Netherlands Astronomical Satellite (NAS) 92
Neutral atoms 86, 94
Neutral molecules 86
Neutron stars 92, 164, 165
Newall Telescope 38
Newcomb, S 28, 69
New General Catalogue 69
New Star in Ophiuchus 18, 65
Newton, Professor HA 33, 45
Newton, Sir Isaac 18, 19, 20-22, 28, 36
Newton's Three Laws of Motion 21, 22
Isaac Newton Telescope, Spain 54, 184
NGC 1068 (galaxy) *176*
NGC 1275 (galaxy) 173, *174*
NGC 1432 (galaxy) *59*, *189*
NGC 1976 (galaxy) *63*
NGC 2440 nucleus *175*
NGC 4261 (galaxy) *176*
NGC 5364 (galaxy) *28*
NGC 6514 (galaxy) *1*, *58*
NGC 6523 (galaxy) *66-67*
NGC 6611 (galaxy) *63*
NGC 7457 (galaxy) *172*
Nice Observatory, France 41, 42, 43
Nicholson, SB 129, 181
Nile River (Earth) 115
1979 J1 *see* Adrastea (moon of Jupiter)
1979 J2 *see* Thebe (moon of Jupiter)
1980 S26 *see* Janus (moon of Saturn)
1981 S13 (moon of Saturn) 137
1989N1 (Proteus) (moon of Neptune) 122, 183
1989N2 (Larissa) (moon of Neptune) 183
1989N3 (Despina) (moon of Neptune) 183
1989N4 (Galatea) (moon of Neptune) 183
1989N5 (Thalassa) (moon of Neptune) 183
1989N6 (Naiad) (moon of Neptune) 183
Nobel Prize 158, 161
Norse mythology 131
North Atlantic Ocean (Earth) 111
North Star *see* Polaris
Northumberland telescope 28
Nova Aurigas (star) 60
Novae 60, 64, *64*, 65, 89, 90, 92, 165, 172, 173, 174
Nova of 1604 (star) 65
Nova Persei (star) 65
Nova Serpentis (star) 89
Nuclear explosions (man-made) 83
Nuclear reactions (natural) 92
Oberon (moon of Uranus) 25, *146*, 183
Observatories of the Carnegie Institute, USA 175
Observatorio Roque de los Muchachos, Spain 54, 184
O'Dell, C Roberts 178
Olbers, Heinrich 32, 33, 45
Olivier, CP 31
Olympus Mons (Mars) 96, *97*
Omega Centauri (star) *161*
Ophelia (moon of Uranus) 147, 182
Ophiuchus (constellation) 18, 65, 69
Optical Astronomy 50-57
Optical Telescope Assembly (OTA) 166

Optical Telescopes *3*, 50-57, *54*, 176
Optics 18
Optics or *A Treatise of the Reflexions, Refractions, Inflexions, and Colors of Light* 21, 22
Orbiting Astronomical Observatories (OAO) 89
    OAO 2 *Stargazer* 89, 94
    OAO 3 *Copernicus* 89, 90
Orbiting Geophysical Observatories (OGO) 92-94, *92*
Orbiting Observatories 158-165, 166
Orbiting Solar Observatories 80
    OSO 1 satellite 80
    OSO 3 satellite 80
    OSO 5 satellite 80
    OSO 8 satellite *81*
Orion (constellation) 9, 24, *63*, *91*
Orion, Great Nebula in 24, 60, 169, *174*, 178, 179
Outer Solar System Missions 116
Outlines of Astronomy 33
Ovda Regio (Venus) 112
Owens Valley Radio Telescope, USA *56*, 185
Oxford University, UK 58
Palermo Observatory, Italy 32, 34
Pallas (asteroid) 32
Palomar Observatory, USA 48, *51*, 53, 54, 57, 65, 129, 184
Pandora (moon of Saturn) 137, 181
Parallactic displacement 15
Parallax 18, 24, 27, 42
Paris Observatory, France 23, 30, 38, 43
Parkes Radio Telescope, Australia 185
Pasiphae (moon of Jupiter) 129, 181
John Paul II (Pope) 20
Pavonis Mons (Mars) *107*
P Cygni (star) 65
Pele (Io) 131
Pendulum clock 20
Penzias, Arno 161
Perkin-Elmer *see* Hughes Danbury Optical Systems
Perrine, CD 45, 129, 181
Perseid meteor showers 33, 46
Perseus (constellation) 46, 65
Peters, CHF 42
Phainomena 8
Philosophie Nauralis Principia Mathematics (The Principia) 21, 22
Phobos (moon of Mars) 40, 41, 94, 180
Phoebe (moon of Saturn) 43, 119, 129, 136, 182
Photochemical processes 88
Photoelectric cell 47
Photographic telescope 39, 47, 60, 65
Photography 27, 28, 30, 32, 36, 38, 39, 42, 43, 45, 46, 47, 48, 50, 58, 60, 61, 62, 69, 73, 74, 76, 92, 98, 101, 103, 104, 109, 119, 121, 129, 131, 136, 137, 138, 147, 148, 166, 169, 172, 176
Photometers 47, 58, 62, 158
Photopolarimeter 146
Piazzi, Giuseppe 27, 32, 34
Pic du Midi, France 47
Pickering, EC 58, 61, 62
Pickering, William Henry 30, 41, 43, 48, 60, 182
Pictor (constellation) 161
Pilachowski, Dr Catherine 62
Pioneer spacecraft 57, 80, 116, 150
    Pioneer 4 spacecraft 80
    Pioneer 6 spacecraft 80
    Pioneer 7 spacecraft 80
    Pioneer 8 spacecraft 80
    Pioneer 9 spacecraft 80
    Pioneer 10 spacecraft 80, 98, *98*, 101, 116
    Pioneer 11 spacecraft 57, *99*, *100*, 101, 116, 136
Pioneer Venus project 109, 112
Pioneer Venus Radar Mapper 112
PKS0528 + 134 (quasar) 161
Max Planck Institite for Radio Astronomy, Germany 185
Planetary Astronomy 38-43
Planetary systems 90, 91, 178-179
Planets 8, 9, 11, 13, 17, 18, 19, 33, 36, 38, 39, 82, 98, 101, 180
Plato 11, 12, *16*
Pleiades 18, 60
Pleiades Nebula (star cluster) *59*, 60, *189*
Pluto 32, 48-49, *48-49*, 80, 98, 101, 119, 169, 180
    Moon of 49, 169
Pogson, N 58, 62
Pogson's Ratio 58
Polaris (star) 6, *7*, 27, 58
Pole Star *see* Polaris

Pollux (star) 36
Pons, Jean Louis 28, 32, 45
Portia (moon of Uranus) 182
Portia (star) 19
Poseidon 29
Potsdamer Photometrische Duchmusterung 58
Potsdam Telescope, Germany 185
Praesepe (star) 18
Prague, Austria 38
Precession of the Equinoxes 13, 17
Prime vertical instrument 39
Princeton Observatory, USA 43
Prismatic camera 60, 65
Prisms 21, 36, 60
Pritchard, C 58
Proctor, RA 42, 45, 69
Procyon (binary star) 22, 27, 36, 61
Project Voyager *see* Voyager Project
Prometheus (moon of Saturn) 137, 181
Protoplanets 178-179
Ptolemy (Claudius Ptolemaeus) 8, 9, *9*, 11, 13, *13*, 14, 15, *16*, 17, 20
Puck (moon of Uranus) 147, 183
Pulkowa Telescope 38, 40, 45
Pulsars *see* Neutron stars
Purbach, George 14
Pyrolytic Release Experiment 104
Pythagoras 12
Pythagorean Theorem 12
Quasars 85, 86, 90, 92, 161, 168, 172, 178
Q0208-512 (quasar) 161
Radiant points 46
Radiation belt (man-made) 83
Radiation belts (natural) 82, 83, 85, 92, 94, 101, 116
Radio Astronomy 57
Radio Astronomy Explorers *see* Explorer 38 and Explorer 49
Radio galaxy 177-178
Radio waves (natural) 85, 89, 165, 174, 176, 177
Radio telescopes 57, 176, 185
Radius vector 18
Ranyard, AC 31, 76
Reber, Grote 57
Red Planet *see* Mars
Reflecting telescopes 9, 18, 19, 20, 24, 25, *25*, 26, *26*, 28, 29, *29*, 31, 38, 41, 47, 50, 53, 54, 60, 69, 184
    Cassegrain 19, 20, 38
    Crossley 38, 69
    Gregorian 19, 20, 24
    Herchelian *25*, 26
    Lassell 29
    Newtonian 19, 20, 24, 26
    Rosse 38
    Schmidt 47
    Tower 50
Refracting telescopes 19, 20, 24, 29, 30, 31, 38, 40, 41, 43, 45, 50, 53, 61, 76, 185
    Achromatic 19
    Chromatic 20
    Coudé *4*, 30, 43
    Great 37
    Non-achromatic 24
Regiomontanus *see* Muller, Johannes
Reichenbach's Institute, Germany 19, 27
RE 1938-461 (binary star) 165
Retrograde orbits 129
Reversing Layer 78-79
De Revolutionibus Orbium Celestium 14
Rhea (moon of Saturn) *139*, 140-142, *141*, 182
Rhea Mons (Venus) 111
R Hydrae (binary star) 62, 64
Rice University, USA 178
Richstone, Douglas 175
Ritchey-Chretien telescope 166
Rittenhouse, D 41
Roberts, I 60, 69
Roche, Edward 39, 136
Rockefeller family 53
Rockwell 116
Rome 6
Roman Catholic Church 20
Roman mythology 6, 41, 48, 102, 108, 124, 129, 134, 137, 140, 148
Romeo and Juliet 20
Rosalind (moon of Uranus) 182
Royal Astronomical Society 22, 24, 25, 26
Royal Astronomical Society of Canada 53
Royal Observatory, England 19, 23, 50, 61, 62, 74, 129, 185
Réseau photospherique (photospheric network) 76

Emperor Rudolf 17
Russell, HN 40
Sabine, E 74
Sagittarius (constellation) *1*, *58*, 66-67
Sappho (asteroid) 39, 40
Saturn 8, 9, 11, *14*, 20, 22, 24, 25, 32, 33, 39, 42, 43, 46, 89, 98, 101, *104*, 116-121, *117*, 122, *123*, *126*, 129, *134*, *135*, 136-137, *138-139*, 140, 143, *143*, 144, 147, 148, 150, 169, *172*, 180
  Moons of 20, 39, 42, 43, 101, 116, 119, *134*, 137-143, *139-142*, 150, 156, 181-182
  Rings of 20, 24, *39*, 42, 101, 116, 119, 134-136, *135-137*, 138, *138*, *139*, *143*, *172*
Saturn launch vehicles 116
Saunders, Steve 112
Sawyer, Scott 49
Scargle, Jeffrey 138
Scheiner, J 60
Schiaparelli, Giovanni 33, 40, 41, 46, 97, 102
Schmidt, B 47
Schmidt camera 47
Schmidt, Julius 30
Schonfeld, E 46, 62
Schroeter, Johann Hieronymus 26, 30, 32, 40, 41, 42, 97
Schwabe, Heinrich 42, 70, 73
Science magazine 115
Scooter (Neptune) 121, *148*, 149
Scorpio (constellation) 65, *165*
Scorpius (constellation) 92
Scout launch vehicle 86
Search for Extra-Terrestrial Intelligence *see* National Aeronautics & Space Administration (NASA)
Secchi, A 45, 58, 60
See, TJJ 45, 61
Seeliger, H 39, 65, 69
Selenographical Society 30
Selenography *4*, 26, 30
Selenotopographische Fragmenter 26
Selenium cell 47
Self-Test-and-Repair (STAR) computer 116
Serpens 63
Seyfert galaxies 172
Shajn Telescope 184
Shakespeare, William 20
C Donald Shane Telescope 184
Shaya, Edward 176
Short, James 19, 73
Showlater, Mark 137-138
'Siddhantas' 11
Siderial Astronomy 24, 26
Sideral period 11
Siding Spring Observatory 53
Sigma Sagittarii (star) 146
Sinope (moon of Jupiter) 129, 181
Sirius (binary star) 20, 22, 27, 36, 60, 61
61 Cygni (binary star) 27, 58
Small Astronomy Satellites (SAS) 85, 161 *see also* Explorer satellites
  Small Astronomical Satellite-2 (SAS-2) 164
Smoot, George 161, 163
Societe Astronomique de France 53
Solar *see* Sun
Solar Astronomy 70-81
Solar constant 73
Solar Maximum Mission (SMM) satellite 80
Solar Mesosphere Explorer 79
Solar parallax 23, 42, 73
Solar period 11
Solar physics 74
Solar System 17, 18, 23, 24, 25, 26, 28, 34, 39, 41, 42, 45, 48, 49, 60, 69, 90, 101, 112, 116-149, 178
Solar telescopes 50, *77*, *78*, *79*
Solstices 9, 13
South Tropical Disturbance 124
Soviet Union 109
Space Shuttle 80, 178
  Atlantis 109, *154*, 159
  Challenger 47, 80
  Columbia 160
  Discovery 168, *170*, *171*
Space Telescope Science Institute *see* Association of Universities for Research in Astronomy (AURA)
Spear, Tony 115
Spectral types of stars 60, 187
Spectrohelioscope 47, 73
Spectrometers *81*
Spectroscope 36, *36*, *38*, 40, 58-69, 73, 76, 172
Spectroscopic binaries 62
Spectrum analysis 21, 31, 36, 38, 39, 40, 41, 43, 45, 46, 58-69, 78, 166, 168, 172

Spencer, Herbert 69
Sphinx *10*
Spica (star) 12
Sporer, FEG 70, 74
Sporer's Law 74
Standing Stones, Brittany 9
Starburst galaxy 176
Star clusters 24, 26, 60, 64, 69, 168, 176, 177, 178
Starfish project 83
Stargazer see Orbiting Astronomical Observatory (OAO) 2
Stationary radiants 46
Steinheil, CA 38
De Stella Nova 15
Stellar Astronomy 58-69
Stellar evolution 89, 90
Stellar parallax 24, 27, 39
Stellar photometry 47
Stellar radial velocity 60
Stellar wind 172, 173
Stokes, GC 36
Stone, Edward C 122
Stonehenge, England 9
Stoney, G Johnstone 31, 46
Struve, FGW 27, 58, 61
Struve, O 61
Sun 6, 8, 9, 10, 11, 12, 13, 17, 18, 19, 20, 21, 22, 23, 24, 26, 27, 28, 30, 31, 32, 33, 34, 36, 38, 39, 40, 41, 42, 45, 46, 47, 48, 50, 60, 64, 69, 70-81, *70-71*, *74-75*, 82, 83, 85, 86, 89, 90, 91, 92, 94, 98, 101, 108, *116*, 119, 121, 122, 124, 143, *143*, 148, 150, 152, 165, 172, 173, 174
  Convection zone 75
  Core 75
  Corona 47, 70, 76, 80, 101
  Chromosphere 75, 76, 81
  Cycle 70-79, 80, 86, 101
  Eclipses 8, 9, 70, *71*, 76, *77*, 78
  Eruptions 86
  Flares *70*, 79, 80, 83, 159-161
  Heliopause 122
  Heliosphere 79, 80, 98, 122
  Magnetic field 70, 73, 74, 79, 80, 83, 94, 98, 101
  Particles 92, 101
  Photosphere 75
  Poles 80
  Prominences 47, 70, 73, *75*, 76, 78
  Radiation 73, 79, 83, 85, 86, 159
  Radio waves 85
  Soundwaves 75
  Spectrum 38
  Sunspots *47*, 50, 70-81, *74-75*
  Wind 79, 80, 82, 83, 85, 86, 92, 94, 98, 101, *109*, 122
Sundial 10, 11
Superbubble of gas 92
Superior planets 9
Supernova 1987A *64*
Supernovas *see* Novae
Support Systems Module (SSM) 166
Synnott, Stephen 138
Synodic period
Syrtis Major (Mars) 102
Tabulae Regiomontanae 27
Taylor, GI 39
Taurus (constellation) 34, *59*, 64, 69, *189*
Taurus-Auriga region 90
Taurus Moving Cluster 69
Tebbutt's Comet 45
Telesto (moon of Saturn) 182
Tempel's Comet 45
Termination shock 122
Terzan 2 (globular star cluster) 92
Tethys (moon of Saturn) *139*, 140, *142*, 182
Thales 11
Tharsis Ridge (Mars) *107*
Thebe (moon of Jupiter) 129, 181
Theia Mons (Venus) 111
Theoria Motus 26
Thermoelectric Outer Planets Spacecraft (TOPS) 116
Theta Orionis (multiple star) 60
30 Doradus (star cluster) 168, *177*
3C279 (quasar) *159*, 161
Three Laws of Motion *see* Newton's Laws
Three Laws of Planetary Motion *see* Kepler's Laws
Tibetan plateau (Earth) 111
Timocharis 12
Tisserand, FF 45
Titan (moon of Saturn) 20, 116, 119, 135, 136, *139*, 142, *142*, 150, *156*, 182

Titan-Centaur launch vehicles 103, 104
Titan launch vehicles 103, 104, 116
Titania (moon of Uranus) 25, *146*, 183
Titius, JD 32
Tombaugh, Clyde 32, 48
Tonry, John 31
Toulouse Telescope, France 38
Tower Telescope 50
Transits
Trifid Nebula *1*, *58*
Trigonometry 12
Triton (moon of Neptune) 29, 49, 122, 129, 136, 149, 183
Tropical Year 13
Troughton, Edward 19
TRW 116
T Tauri (star) 91
T Tauri stars 91
Uhuru *see* Explorer 42 satellite
Ultraviolet 50, 80, 85, 89, 92, 94, 98, 122, 124, 143, 165, 166, 168, 169, 173
Umbriel (moon of Uranus) *146*, 183
United Kingdom 86, 90
United Kingdom Infrared Telescope (UKIRT), USA 54, 184
US Department of Defense 57
US Naval Observatory, USA 41, 45, 49, 50
US Navy 83
University Explorer *see* Explorer 40 satellite
University of Arizona, USA 121
University of California, USA 57, 164, 175
University of Chicago, USA 38, 50, 53
University of Colorado, USA 79, 173
University of Hawaii's Institute for Astronomy, USA 54
University of Hawaii Telescope 184
University of Iowa, USA 86, 101
University of Maryland, USA 176
University of Michigan, USA 175
Uranometria Argentina 58
Uranometria Nova Oxoniensis 58
Uranus 25, 26, 28, 29, 32, 33, 43, 45, 48, 116-121, 122, *134*, 138, 144-147, *144-145*, *146*, *147*, 148, 180
  Moons of 43, 45, 121, 146-147, 182-183
  Rings of 144, 147, *147*
Ursa Major (constellation) 6, *160*, 161
Utopia Planitia (Mars) *107*
UY Aurigae (star) 91
Valhalla (Callisto) 133, *133*
Valles Marinaris (Mars) 96
Van Allen, James 101
Van Allen Radiation Region 82, 83, 85, 92, 94, 101
Variable stars 24, 47, 62, 64, 166
Variation of latitude 45
Variation of the Compass 22
Vega (star) 27, 38, 58, 60, 90
Vela Nebula 164, 165
Venera spacecraft 109, 180
  Venera 9 spacecraft 109
  Venera 10 spacecraft 109
  Venera 13 spacecraft 109
  Venera 14 spacecraft 109
Venus 8, 9, 10, 11, 12, 17, 18, 19, 22, 23, 25, 33, 36, 39, 40, 41, 94, 96, *96*, 98, 108-115, *108*, *109*, *111*, 142, 148, 150, 156, 180
  Transits of 23, 39
Vermillian River Radio Telescope, USA 185
Very, FW 31
Very Large Array, USA 56, 57, 185
Vesta (asteroid) 32
Victoria (asteroid) 39, 40
Victoria Telescope, Canada 50
Vienna Telescope, Austria 23, 38, 42, 185
Viking Mars Orbiters 102-107, *102-107*, 119, 150, 180
  Viking 1 spacecraft 103
  Viking 2 spacecraft 103-104
Virgo (constellation) 12, 28, *28*, 161, 172, 174, *178*
Vogel, HC 41, 60, 61
von Humboldt, A 33
Von Madler, Johann Heinrich 30
Voyager Project 116-149, *116-119*, 121, *125*, *143*, 149
  Spacecraft 43, 45, 101, 116-149, *143*, 150, 180, 181, 182, 183
  Voyager Interstellar Mission (VIM) 122
  Voyager 1 spacecraft 116-149, *122*, *123*, 128, 129, *135*
  Voyager 2 spacecraft 101, 116, *116*, 119-149, *128*

WD1620-391 (binary star system) *165*
Wedge photometer 58
Weiler, Ed 165
Weinek, L 30
Wells' Comet 45
Fred Lawrence Whipple Observatory, USA 54, *54*, 184
Whirlpool Galaxy *see* M51
Wide Field/Planetary Camera (WFPC) 166, 169, 172, 175, 176
Wide Field/Planetary Camera-2 (WFPC-2) 169, *172*, *178*
Williams, Stanley 43
Wilson, Robert 161
Wilson-Herschel Theory 70
Wind shear 88
Winnecke's Comet 33, 45
Wolf, Max 42, 60, 65, 69
Wolf, R 74
Wolf-Rayet stars 60
Wollaston, WH 36
World War II 32, 53, 57
Wright of Durham 39
Xi Geminorum (variable star) 64
X-ray bursters 92
X-rays 80, 82, 85, 89, 92, 93, *93*, 161, 164, 165, 174
Yale College Observatory, USA 46
Yale University, USA 33
Yerkes, Charles Tyson 53
Yerkes Telescope 38, 45, 61, 65, 185
Yerkes Telescope, USA 50, 185
Young, CA 43, 76, 78
Yu Shih 10
Zenith Sector 19, *23*
0537-441 (BL Lacertae object) 161
0717 + 714 (BL Lacertae object) 161
Zeta Ursae Majoris (binary star) 60, 61, 62
Zodiac 8, 11, 25, 135
Zodiacal Light 20
Zollner, JKF 42, 45, 60, 76
Zurich University Observatory, Switzerland 47

PICTURE CREDITS

All photos appear through the courtesy of the National Aeronautics and Space Administration with the following exceptions:
AGS Archives: 8, 9, 12, 13, 14 (both), 16, 17 (both), 18 (both), 19, 20, 21 (all), 23 (both), 25 (both), 27, 29, 31, 36, 38, 40 (bottom), 42, 43, 50
California Institute of Technology: 33 (right), 51, 52
Hale Observatories: 1, 59, 63 (bottom), 103 (bottom)
Harvard College Observatory: 37
Hughes Aircraft Company: 81
Mike Jensen/Australian Information Service: 53
Kitt Peak National Observatory: 28, 58, 63 (top), 66-67, 77 (both)
Lowell Observatory: 49 (top, left & right)
© Reverend MJ McPike Collection: 15
National Optical Astronomy Observatories: 2, 3, 4, 6, 7, 54 (top), 55, 62, 64 (both), 68, 71, 72, 73, 74 (both), 75 (top), 76, 78-79 (all), 189
National Radio Astronomy Observatory/AUI: 56 (both), 57
Smithsonian Institution: 40 (top)
© Stock Editions: 10, 11
US Naval Observatory: 41, 69
Whipple Observatory, Smithsonian Institution: 54 (bottom)
Yale University, Department of Astronomy: 44
© Bill Yenne: 48-49 (bottom), 126 (both)